WATCH AND CLOCK ESCAPEMENTS

A Complete Study In Theory and Practice of the Lever, Cylinder and Chronometer Escapements, Together with a Brief Account of the Origin and Evolution of the Escapement in Horology

Compiled from the well-known Escapement Serials published in The Keystone

NEARLY TWO HUNDRED ORIGINAL ILLUSTRATIONS

PREFACE

Especially notable among the achievements of The Keystone in the field of horology were the three serials devoted to the lever, cylinder and chronometer escapements. So highly valued were these serials when published that on the completion of each we were importuned to republish it in book form, but we deemed it advisable to postpone such publication until the completion of all three, in order that the volume should be a complete treatise on the several escapements in use in horology. The recent completion of the third serial gave us the opportunity to republish in book form, and the present volume is the result. We present it to the trade and students of horology happy in the knowledge that its contents have already received their approval. An interesting addition to the book is the illustrated story of the escapements, from the first crude conceptions to their present perfection.

Contents

PREFACE i
I. THE DETACHED LEVER ESCAPEMENT 1
II. THE CYLINDER ESCAPEMENT 132
III. THE CHRONOMETER ESCAPEMENT 155
IV. HISTORY OF ESCAPEMENTS...............................176
V. PUTTING IN A NEW CYLINDER 201

I. THE DETACHED LEVER ESCAPEMENT

In this treatise we do not propose to go into the history of this escapement and give a long dissertation on its origin and evolution, but shall confine ourselves strictly to the designing and construction as employed in our best watches. By designing, we mean giving full instructions for drawing an escapement of this kind to the best proportions. The workman will need but few drawing instruments, and a drawing-board about 15" by 18" will be quite large enough. The necessary drawing-instruments are a T-square with 15" blade; a scale of inches divided into decimal parts; two pairs dividers with pen and pencil points—one pair of these dividers to be 5" and the other 6"; one ruling pen. Other instruments can be added as the workman finds he needs them. Those enumerated above, however, will be all that are absolutely necessary.

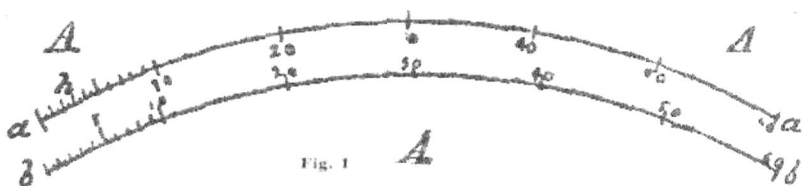

Fig. 1

We shall, in addition, need an arc of degrees, which we can best make for ourselves. To construct one, we procure a piece of No. 24 brass, about 5-1/2" long by 1-1/4" wide. We show such a piece of brass at A, Fig. 1. On this piece of brass we sweep two arcs with a pair of dividers set at precisely 5", as shown (reduced) at $a\ a$ and $b\ b$. On these arcs we set off the space held in our dividers—that is 5"—as shown at the short radial lines at each end of the two arcs. Now it is a well-known fact that the space embraced by our dividers contains exactly sixty degrees of the arcs $a\ a$ and $b\ b$, or one-sixth of the entire circle; consequently, we divide the arcs *a a* and *b b* into sixty equal parts, to represent degrees, and at one end of these arcs we halve five spaces so we can get at half degrees.

Fig 2

Before we take up the details of drawing an escapement we will say a few words about "degrees," as this seems to be something difficult to understand by most pupils in horology when learning to draw parts of watches to scale. At Fig. 2 we show several short arcs of fifteen degrees, all having the common center *g*. Most learners seem to have an idea that a degree must be a specific space, like an inch or a foot. Now the first thing in learning to draw an escapement is to fix in our minds the fact that the extent of a degree depends entirely on the radius of the arc we employ. To aid in this explanation we refer to Fig. 2. Here the arcs *c*, *d*, *e* and *f* are all fifteen degrees, although the linear extent of the degree on the arc *c* is twice that of the degree on the arc *f*. When we speak of a degree in connection with a circle we mean the one-three-hundred-and-sixtieth part of the periphery of such a circle. In dividing the arcs *a a* and *b b* we first divide them into six spaces, as shown, and each of these spaces into ten minor spaces, as is also shown. We halve five of the degree spaces, as shown at *h*. We should be very careful about making the degree arcs shown at Fig. 1, as the accuracy of our drawings depends a great deal on the perfection of the division on the scale *A*. In connection with such a fixed scale of degrees as is shown at Fig. 1, a pair of small dividers, constantly set to a degree space, is very convenient.

2

MAKING A PAIR OF DIVIDERS.

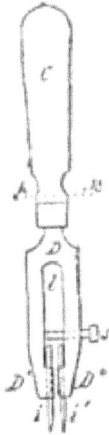

Fig. 3

To make such a pair of small dividers, take a piece of hard sheet brass about 1/20" thick, 1/4" wide, 1-1/2" long, and shape it as shown at Fig. 3. It should be explained, the part cut from the sheet brass is shown below the dotted line k, the portion above (C) being a round handle turned from hard wood or ivory. The slot l is sawn in, and two holes drilled in the end to insert the needle points i i. In making the slot l we arrange to have the needle points come a little too close together to agree with the degree spaces on the arcs a a and b b. We then put the small screw j through one of the legs D'', and by turning j, set the needle points i i to exactly agree with the degree spaces. As soon as the points i i are set correctly, j should be soft soldered fast.

The degree spaces on A are set off with these dividers and the spaces on A very carefully marked. The upper and outer arc a a should have the spaces cut with a graver line, while the lower one, b b is best permanently marked with a carefully-made prick punch. After the arc a a is divided, the brass plate A is cut back to this arc so the divisions we have just made are on the edge. The object of having two arcs on the plate A is, if we desire to get at the number of degrees contained in any arc of a 5" radius we lay the scale A so the edge agrees with the arc a a, and read off the number of degrees from the scale. In setting dividers we employ the dotted spaces on the arc b b.

3

DELINEATING AN ESCAPE WHEEL.

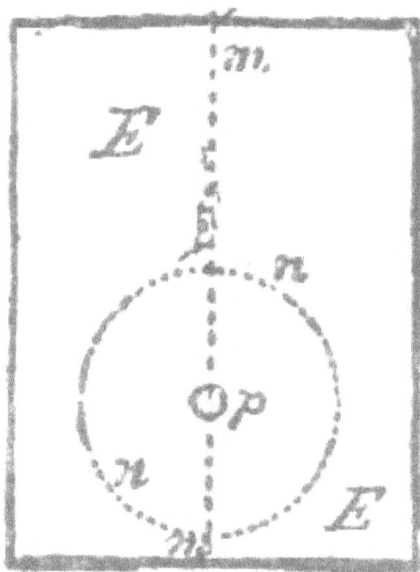

Fig. 4

We will now proceed to delineate an escape wheel for a detached lever. We place a piece of good drawing-paper on our drawing-board and provide ourselves with a very hard (HHH) drawing-pencil and a bottle of liquid India ink. After placing our paper on the board, we draw, with the aid of our T-square, a line through the center of the paper, as shown at *m m*, Fig. 4. At 5-1/2" from the lower margin of the paper we establish the point *p* and sweep the circle *n n* with a radius of 5". We have said nothing about stretching our paper on the drawing-board; still, carefully-stretched paper is an important part of nice and correct drawing. We shall subsequently give directions for properly stretching paper, but for the present we will suppose the paper we are using is nicely tacked to the face of the drawing-board with the smallest tacks we can procure. The paper should not come quite to the edge of the drawing-board, so as to interfere with the head of the T-square. We are now ready to commence delineating our escape wheel and a set of pallets to match.

4

WATCH AND CLOCK ESCAPEMENTS

The simplest form of the detached lever escapement in use is the one known as the "ratchet-tooth lever escapement," and generally found in English lever watches. This form of escapement gives excellent results when well made; and we can only account for it not being in more general use from the fact that the escape-wheel teeth are not so strong and capable of resisting careless usage as the club-tooth escape wheel.

It will be our aim to convey broad ideas and inculcate general principles, rather than to give specific instructions for doing "one thing one way." The ratchet-tooth lever escapements of later dates have almost invariably been constructed on the ten-degree lever-and-pallet-action plan; that is, the fork and pallets were intended to act through this arc. Some of the other specimens of this escapement have larger arcs—some as high as twelve degrees.

PALLET-AND-FORK ACTION.

Fig. 5

We illustrate at Fig. 5 what we mean by ten degrees of pallet-and-fork action. If we draw a line through the center of the pallet staff, and also through the center of the fork slot, as shown at $a\,b$, Fig. 5, and allow the fork to vibrate five degrees each side of said lines $a\,b$, to the lines $a\,c$ and $a\,c'$, the fork has what we term ten-degree pallet action. If the fork and pallets vibrate six degrees on each side of the line $a\,b$—that is, to the lines $a\,d$ and $a\,d'$—we have twelve degrees pallet action. If we cut the arc down so the oscillation is only four and one-quarter degrees on each side of $a\,b$, as indicated by the lines $a\,s$ and $a\,s'$, we have a pallet-and-fork action of eight and one-half degrees; which, by the way, is a very desirable arc for a carefully-constructed escapement.

The controlling idea which would seem to rule in constructing a detached lever escapement, would be to make it so the balance is

5

free of the fork; that is, detached, during as much of the arc of the vibration of the balance as possible, and yet have the action thoroughly sound and secure. Where a ratchet-tooth escapement is thoroughly well-made of eight and one-half degrees of pallet-and-fork action, ten and one-half degrees of escape-wheel action can be utilized, as will be explained later on.

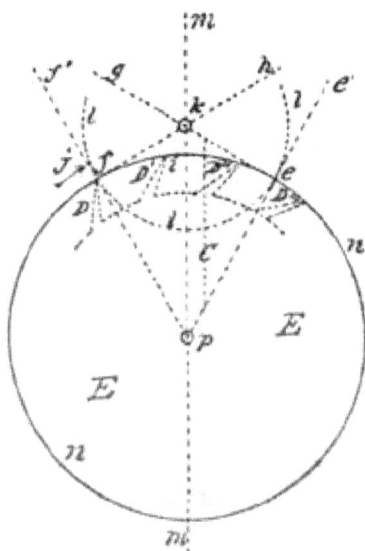

Fig. 6

We will now resume the drawing of our escape wheel, as illustrated at Fig. 4. In the drawing at Fig. 6 we show the circle *n n*, which represents the periphery of our escape wheel; and in the drawing we are supposed to be drawing it ten inches in diameter.

We produce the vertical line *m* passing through the center *p* of the circle *n*. From the intersection of the circle *n* with the line *m* at *i* we lay off thirty degrees on each side, and establish the points *e f*; and from the center *p*, through these points, draw the radial lines *p e'* and *p f'*. The points *f e*, Fig. 6, are, of course, just sixty degrees apart and represent the extent of two and one-half teeth of the escape wheel. There are two systems on which pallets for lever escapements are made, viz., equidistant lockings and circular pallets. The advantages claimed for each system will be discussed subsequently. For the first and present illustration we will assume we are to employ circular pal-

6

lets and one of the teeth of the escape wheel resting on the pallet at the point f; and the escape wheel turning in the direction of the arrow j. If we imagine a tooth as indicated at the dotted outline at D, Fig. 6, pressing against a surface which coincides with the radial line $p\,f$, the action would be in the direction of the line $f\,h$ and at right angles to $p\,f$. If we reason on the action of the tooth D, as it presses against a pallet placed at f, we see the action is neutral.

ESTABLISHING THE CENTER OF PALLET STAFF.

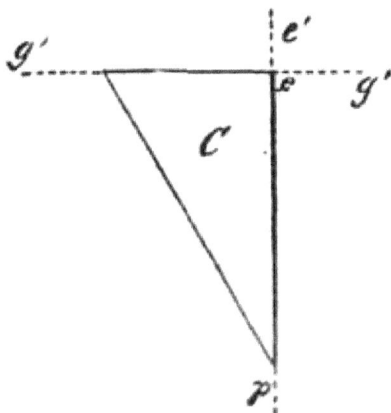

Fig. 7

      With a fifteen-tooth escape wheel each tooth occupies twenty-four degrees, and from the point f to e would be two and one-half tooth-spaces. We show the dotted points of four teeth at $D\ D'\ D''\ D'''$. To establish the center of the pallet staff we draw a line at right angles to the line $p\,e'$ from the point e so it intersects the line $f\,h$ at k. For drawing a line at right angles to another line, as we have just done, a hard-rubber triangle, shaped as shown at C, Fig. 7, can be employed. To use such a triangle, we place it so the right, or ninety-degrees angle, rests at e, as shown at the dotted triangle C, Fig. 6, and the long side coincides with the radial line $p\,e'$. If the short side of the hard-rubber triangle is too short, as indicated, we place a short ruler so it rests against the edge, as shown at the dotted line $g\,e$, Fig. 7, and while holding it securely down on the drawing we remove the triangle, and with a fine-pointed pencil draw the line $e\,g$, Fig. 6, by the short rule. Let us imagine a flat surface placed at e so its face was at right angles

7

to the line *g e*, which would arrest the tooth *D″* after the tooth *D* resting on *f* had been released and passed through an arc of twelve degrees. A tooth resting on a flat surface, as imagined above, would also rest dead. As stated previously, the pallets we are considering have equidistant locking faces and correspond to the arc *l l*, Fig. 6.

In order to realize any power from our escape-wheel tooth, we must provide an impulse face to the pallets faced at *f e*; and the problem before us is to delineate these pallets so that the lever will be propelled through an arc of eight and one-half degrees, while the escape wheel is moving through an arc of ten and one-half degrees. We make the arc of fork action eight and one-half degrees for two reasons—(1) because most text-books have selected ten degrees of fork-and-pallet action; (2) because most of the finer lever escapements of recent construction have a lever action of less than ten degrees.

LAYING OUT ESCAPE-WHEEL TEETH.

To "lay out" or delineate our escape-wheel teeth, we continue our drawing shown at Fig. 6, and reproduce this cut very nearly at Fig. 8. With our dividers set at five inches, we sweep the short arc *a a′* from *f* as a center. It is to be borne in mind that at the point *f* is located the extreme point of an escape-wheel tooth. On the arc *a a* we lay off from *p* twenty-four degrees, and establish the point *b*; at twelve degrees beyond *b* we establish the point *c*. From *f* we draw the lines *f b* and *f c*; these lines establishing the form and thickness of the tooth *D*. To get the length of the tooth, we take in our dividers one-half a tooth space, and on the radial line *p f* establish the point *d* and draw circle *d′ d′*.

To facilitate the drawing of the other teeth, we draw the circles *d′ c′*, to which the lines *f b* and *f c* are tangent, as shown. We divide the circle *n n*, representing the periphery of our escape wheel, into fifteen spaces, to represent teeth, commencing at *f* and continued as shown at *o o* until the entire wheel is divided. We only show four teeth complete, but the same methods as produced these will produce them all. To briefly recapitulate the instructions for drawing the teeth for the ratchet-tooth lever escapement: We draw the face of the teeth at an angle of twenty-four degrees to a radial line; the back of the tooth at an

8

angle of thirty-six degrees to the same radial line; and make teeth half a tooth-space deep or long.

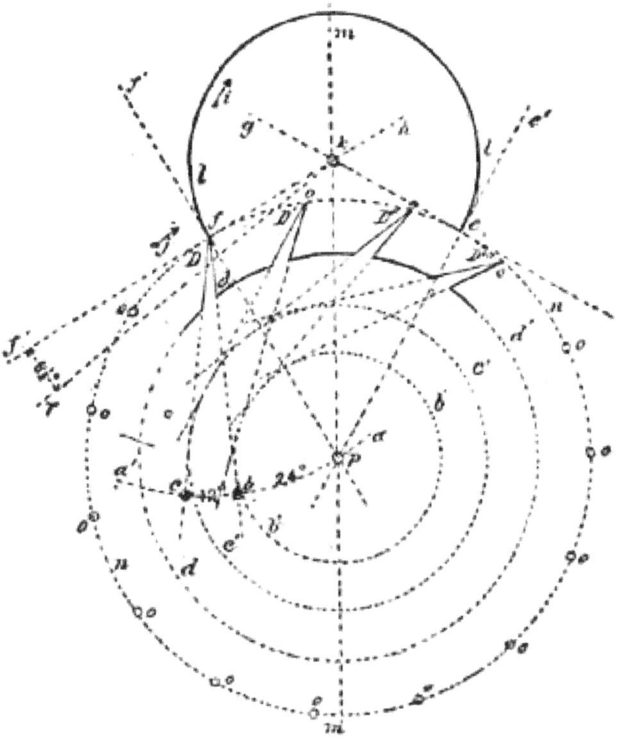

Fig. 8

We now come to the consideration of the pallets and how to delineate them. To this we shall add a careful analysis of their action. Let us, before proceeding further, "think a little" over some of the factors involved. To aid in this thinking or reasoning on the matter, let us draw the heavy arc *l* extending from a little inside of the circle *n* at *f* to the circle *n* at *e*. If now we imagine our escape wheel to be pressed forward in the direction of the arrow *j*, the tooth *D* would press on the arc *l* and be held. If, however, we should revolve the arc *l* on the center *k* in the direction of the arrow *i*, the tooth *D* would *escape* from the edge of *l* and the tooth *D″* would pass through an arc (reckoning from the center *p*) of twelve degrees, and be arrested by the inside of the arc *l* at *e*. If we now should reverse the motion and turn the arc *l* backward, the tooth at *e* would, in turn, be released and the tooth following

9

after D (but not shown) would engage l at f. By supplying motive to revolve the escape wheel (E) represented by the circle n, and causing the arc l to oscillate back and forth in exact intervals of time, we should have, in effect, a perfect escapement. To accomplish automatically such oscillations is the problem we have now on hand.

HOW MOTION IS OBTAINED.

In clocks, the back-and-forth movement, or oscillating motion, is obtained by employing a pendulum; in a movable timepiece we make use of an equally-poised wheel of some weight on a pivoted axle, which device we term a balance; the vibrations or oscillations being obtained by applying a coiled spring, which was first called a "pendulum spring," then a "balance spring," and finally, from its diminutive size and coil form, a "hairspring." We are all aware that for the motive power for keeping up the oscillations of the escaping circle l we must contrive to employ power derived from the teeth D of the escape wheel. About the most available means of conveying power from the escape wheel to the oscillating arc l is to provide the lip of said arc with an inclined plane, along which the tooth which is disengaged from l at f to slide and move said arc l through—in the present instance an arc of eight and one-half degrees, during the time the tooth D is passing through ten and one-half degrees. This angular motion of the arc l is represented by the radial lines $k\,f'$ and $k\,r$, Fig. 8. We desire to impress on the reader's mind the idea that each of these angular motions is not only required to be made, but the motion of one mobile must convey power to another mobile.

In this case the power conveyed from the mainspring to the escape wheel is to be conveyed to the lever, and by the lever transmitted to the balance. We know it is the usual plan adopted by text-books to lay down a certain formula for drawing an escapement, leaving the pupil to work and reason out the principles involved in the action. In the plan we have adopted we propose to induct the reader into the why and how, and point out to him the rules and methods of analysis of the problem, so that he can, if required, calculate mathematically exactly how many grains of force the fork exerts on the jewel pin, and also how much (or, rather, what percentage) of the motive power is lost in various "power leaks," like "drop" and lost motion. In the present case the mechanical result we desire to obtain is to cause our lever pivoted at *k* to vibrate back and forth through an arc of eight and one-half degrees; this lever not only to vibrate back and forth, but also to lock and hold the escape wheel during a certain period of time; that is, through the period of time the balance is performing its excursion and the jewel pin free and detached from the fork.

We have spoken of paper being employed for drawings, but for very accurate delineations we would recommend the horological student to make drawings on a flat metal plate, after perfectly smoothing the surface and blackening it by oxidizing.

PALLET-AND-FORK ACTION.

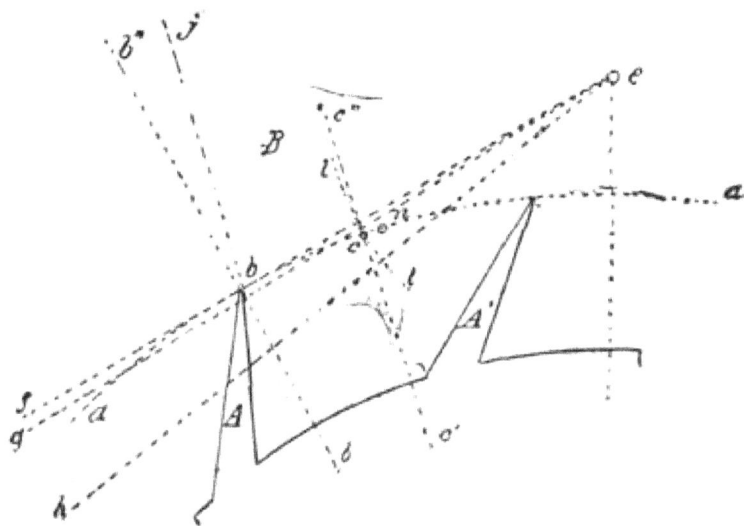

Fig. 9

By adopting eight and one-half degrees pallet-and-fork action we can utilize ten and one-half degrees of escape-wheel action. We show at *A A'*, Fig. 9, two teeth of a ratchet-tooth escape wheel reduced one-half; that is, the original drawing was made for an escape wheel ten inches in diameter. We shall make a radical departure from the usual practice in making cuts on an enlarged scale, for only such parts as we are talking about. To explain, we show at Fig. 10 about one-half

11

of an escape wheel one eighth the size of our large drawing; and when we wish to show some portion of such drawing on a larger scale we will designate such enlargement by saying one-fourth, one-half or full size.

At Fig. 9 we show at half size that portion of our escapement embraced by the dotted lines *d*, Fig. 10. This plan enables us to show very minutely such parts as we have under consideration, and yet occupy but little space. The arc *a*, Fig. 9, represents the periphery of the escape wheel. On this line, ten and one-half degrees from the point of the tooth *A*, we establish the point *c* and draw the radial line *c c'*. It is to be borne in mind that the arc embraced between the points *b* and *c* represents the duration of contact between the tooth *A* and the entrance pallet of the lever. The space or short arc *c n* represents the "drop" of the tooth.

This arc of one and one-half degrees of escape-wheel movement is a complete loss of six and one-fourth per cent. of the entire power of the mainspring, as brought down to the escapement; still, up to the present time, no remedy has been devised to overcome it. All the other escapements, including the chronometer, duplex and cylinder, are quite as wasteful of power, if not more so. It is usual to construct ratchet-tooth pallets so as to utilize but ten degrees of escape-wheel action; but we shall show that half a degree more can be utilized by adopting the eight and one-half degree fork action and employing a double-roller safety action to prevent over-banking.

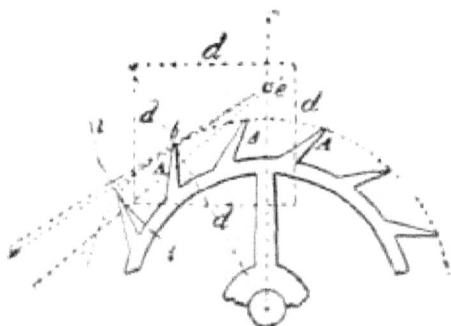

Fig. 10

12

From the point e, which represents the center of the pallet staff, we draw through b the line $e f$. At one degree below $e f$ we draw the line $e g$, and seven and one-half degrees below the line $e g$ we draw the line $e h$. For delineating the lines $e g$, etc., correctly, we employ a degree-arc; that is, on the large drawing we are making we first draw the line $e b f$, Fig. 10, and then, with our dividers set at five inches, sweep the short arc i, and on this lay off first one degree from the intersection of $f e$ with the arc i, and through this point draw the line $e g$.

From the intersection of the line $f e$ with the arc i we lay off eight and one-half degrees, and through this point draw the line $e h$. Bear in mind that we are drawing the pallet at B to represent one with eight and one-half degrees fork-and-pallet action, and with equidistant lockings. If we reason on the matter under consideration, we will see the tooth A and the pallet B, against which it acts, part or separate when the tooth arrives at the point c; that is, after the escape wheel has moved through ten and one-half degrees of angular motion, the tooth drops from the impulse face of the pallet and falls through one and one-half degrees of arc, when the tooth A'', Fig. 10, is arrested by the exit pallet.

To locate the position of the inner angle of the pallet B, sweep the short arc l by setting the dividers so one point or leg rests at the center e and the other at the point c. Somewhere on this arc l is to be located the inner angle of our pallet. In delineating this angle, Moritz Grossman, in his "Prize Essay on the Detached Lever Escapement," makes an error, in Plate III of large English edition, of more than his entire lock, or about two degrees. We make no apologies for calling attention to this mistake on the part of an authority holding so high a position on such matters as Mr. Grossman, because a mistake is a mistake, no matter who makes it.

We will say no more of this error at present, but will farther on show drawings of Mr. Grossman's faulty method, and also the correct method of drawing such a pallet. To delineate the locking face of our pallet, from the point formed by the intersection of the lines $e g b b'$, Fig. 9, as a center, we draw the line j at an angle of twelve degrees to b b''. In doing this we employ the same method of establishing the angle as we made use of in drawing the lines $e g$ and $e h$, Fig. 10. The line j

establishes the locking face of the pallet *B*. Setting the locking face of the pallet at twelve degrees has been found in practice to give a safe "draw" to the pallet and keep the lever secure against the bank. It will be remembered the face of the escape-wheel tooth was drawn at twenty-four degrees to a radial line of the escape wheel, which, in this instance, is the line *b b'*, Fig. 9. It will now be seen that the angle of the pallet just halves this angle, and consequently the tooth *A* only rests with its point on the locking face of the pallet. We do not show the outlines of the pallet *B*, because we have not so far pointed out the correct method of delineating it.

METHODS OF MAKING GOOD DRAWING INSTRUMENTS.

Fig. 11

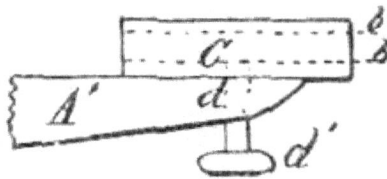

Fig. 12

Perhaps we cannot do our readers a greater favor than to digress from the study of the detached lever escapement long enough to say a few words about drawing instruments and tablets or surfaces on which to delineate, with due precision, mechanical designs or drawings. Ordinary drawing instruments, even of the higher grades, and costing a good deal of money, are far from being satisfactory to a man who has the proper idea of accuracy to be rated as a first-class mechanic. Ordinary compasses are obstinate when we try to set them to the hundredth of an inch; usually the points are dull and ill-shapen; if they make a puncture in the paper it is unsightly.

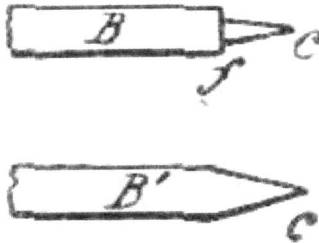

Fig. 13

Watchmakers have one advantage, however, because they can very easily work over a cheap set of drawing instruments and make them even superior to anything they can buy at the art stores. To illustrate, let us take a cheap pair of brass or German-silver five-inch dividers and make them over into needle points and "spring set." To do this the points are cut off at the line *a a*, Fig 11, and a steel tube is gold-soldered on each leg. The steel tube is made by taking a piece of steel wire which will fit a No. 16 chuck of a Whitcomb lathe, and drilling a hole in the end about one-fourth of an inch deep and about the size of a No. 3 sewing needle. We Show at Fig. 12 a view of the point *A'*, Fig. 11, enlarged, and the steel tube we have just drilled out attached at *C*. About the best way to attach *C* is to solder. After the tube *C* is attached a hole is drilled through *A'* at *d*, and the thumb-screw *d* inserted. This thumb-screw should be of steel, and hardened and tempered. The use of this screw is to clamp the needle point. With such a device as the tube *C* and set-screw *d*, a No. 3 needle is used for a point; but for drawings on paper a turned point, as shown at Fig 13, is to be preferred. Such points can be made from a No. 3 needle after softening enough to be turned so as to form the point *c*. This point at the shoulder f should be about 12/1000 of an inch, or the size of a fourth-wheel pivot to an eighteen size movement.

The idea is, when drawing on paper the point c enters the paper. For drawing on metal the form of the point is changed to a simple cone, as shown at B' c, Fig. 13. such cones can be turned carefully, then hardened and tempered to a straw color; and when they become dull, can be ground by placing the points in a wire chuck and dressing them up with an emery buff or an Arkansas slip. The opposite leg of the dividers is the one to which is attached the spring for close setting of the points.

Fig. 14

In making this spring, we take a piece of steel about two and one-fourth inches long and of the same width as the leg of the divider, and attach it to the inside of the leg as shown at Fig. 14, where D represents the spring and A the leg of the dividers. The spring D has a short steel tube C'' and set-screw d'' for a fine point like B or B'. In the lower end of the leg A, Fig. 14, is placed the milled-head screw g, which serves to adjust the two points of the dividers to very close distances. The spring D is, of course, set so it would press close to the leg A if the screw g did not force it away.

SPRING AND ADJUSTING SCREW FOR DRAWING INSTRUMENTS.

It will be seen that we can apply a spring D and adjusting screw opposite to the leg which carries the pen or pencil point of all our dividers if we choose to do so; but it is for metal drawing that such points are of the greatest advantage, as we can secure an accuracy very gratifying to a workman who believes in precision. For drawing circles on metal, "bar compasses" are much the best, as they are almost entirely free from spring, which attends the jointed compass. To make

16

(because they cannot be bought) such an instrument, take a piece of flat steel, one-eighth by three-eighths of an inch and seven inches long, and after turning and smoothing it carefully, make a slide half an inch wide, as shown at Fig. 15, with a set-screw h on top to secure it at any point on the bar E. In the lower part of the slide F is placed a steel tube like C, shown in Figs. 12 and 14, with set-screw for holding points like B B', Fig. 13. At the opposite end of the bar E is placed a looped spring G, which carries a steel tube and point like the spring D, Fig. 14. Above this tube and point, shown at j, Fig. 15, is placed an adjustment screw k for fine adjustment. The inner end of the screw k rests against the end of the bar E. The tendency of the spring G is to close upon the end of E; consequently if we make use of the screw k to force away the lower end of G, we can set the fine point in j to the greatest exactness.

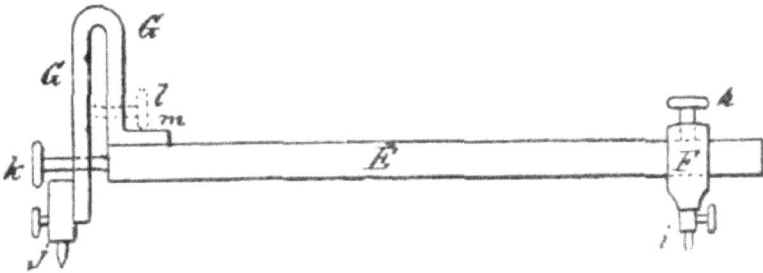

Fig. 15

The spring G is made of a piece of steel one-eighth of an inch square, and secured to the bar E with a screw and steady pins at m. A pen and pencil point attachment can be added to the spring G; but in case this is done it would be better to make another spring like G without the point j, and with the adjusting screw placed at l. In fitting pen and pencil points to a spring like G it would probably be economical to make them outright; that is, make the blades and screw for the ruling pen and a spring or clamping tube for the pencil point.

CONSIDERATION OF DETACHED LEVER
ESCAPEMENT RESUMED.

We will now, with our improved drawing instruments, resume the consideration of the ratchet-tooth lever escapement. We reproduce at Fig. 16 a portion of diagram III, from Moritz Grossmann's "Prize

Essay on the Detached Lever Escapement," in order to point out the error in delineating the entrance pallet to which we previously called attention. The cut, as we give it, is not quite one-half the size of Mr. Grossmann's original plate.

In the cut we give the letters of reference employed the same as on the original engraving, except where we use others in explanation. The angular motion of the lever and pallet action as shown in the cut is ten degrees; but in our drawing, where we only use eight and one-half degrees, the same mistake would give proportionate error if we did not take the means to correct it. The error to which we refer lies in drawing the impulse face of the entrance pallet. The impulse face of this pallet as drawn by Mr. Grossmann would not, from the action of the engaging tooth, carry this pallet through more than eight degrees of angular motion; consequently, the tooth which should lock on the exit pallet would fail to do so, and strike the impulse face.

We would here beg to add that nothing will so much instruct a person desiring to acquire sound ideas on escapements as making a large model. The writer calls to mind a wood model of a lever escapement made by one of the "boys" in the Elgin factory about a year or two after Mr. Grossmann's prize essay was published. It went from hand to hand and did much toward establishing sound ideas as regards the correct action of the lever escapement in that notable concern.

If a horological student should construct a large model on the lines laid down in Mr. Grossmann's work, the entrance pallet would be faulty in form and would not properly perform its functions. Why? perhaps says our reader. In reply let us analyze the action of the tooth B as it rests on the pallet A. Now, if we move this pallet through an angular motion of one and one-half degrees on the center g (which also represents the center of the pallet staff), the tooth B is disengaged from the locking face and commences to slide along the impulse face of the pallet and "drops," that is, falls from the pallet, when the inner angle of the pallet is reached.

This inner angle, as located by Mr. Grossmann, is at the intersection of the short arc i with the line g n, which limits the ten-degree angular motion of the pallets. If we carefully study the drawing, we will see the pallet has only to move through eight degrees of angular

18

motion of the pallet staff for the tooth to escape, *because the tooth certainly must be disengaged when the inner angle of the pallet reaches the peripheral line a.* The true way to locate the position of the inner angle of the pallet, is to measure down on the arc *i* ten degrees from its intersection with the peripheral line *a* and locate a point to which a line is drawn from the intersection of the line *g m* with the radial line *a c*, thus defining the inner angle of the entrance pallet. We will name this point the point *x.*

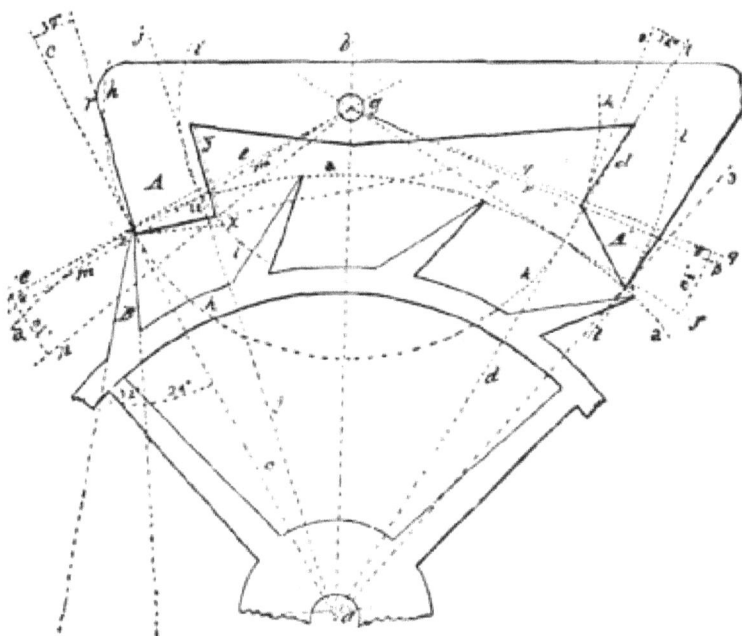

Fig. 16

It may not be amiss to say the arc *i* is swept from the center *g* through the point *u*, said point being located ten degrees from the intersection of the radial *a c* with the peripheral line *a*. It will be noticed that the inner angle of the entrance pallet *A* seems to extend inward, beyond the radial line *a j*, that is, toward the pallet center *g*, and gives the appearance of being much thicker than the exit pallet *A'*; but we will see on examination that the extreme angle *x* of the entrance pallet

19

must move on the arc *i* and, consequently, cross the peripheral line *a* at the point *u*. If we measure the impulse faces of the two pallets *A A'*, we will find them nearly alike in linear extent.

Fig. 17

Mr. Grossmann, in delineating his exit pallet, brings the extreme angle (shown at *4*) down to the periphery of the escape, as shown in the drawing, where it extends beyond the intersection of the line *g f* with the radial line *a 3*. The correct form for the entrance pallet should be to the dotted line *z x y*.

We have spoken of engaging and disengaging frictions; we do not know how we can better explain this term than by illustrating the idea with a grindstone. Suppose two men are grinding on the same stone; each has, say, a cold chisel to grind, as shown at Fig. 17, where *G* represents the grindstone and *N N'* the cold chisels. The grindstone is supposed to be revolving in the direction of the arrow. The chisels *N* and *N'* are both being ground, but the chisel *N'* is being cut much the more rapidly, as each particle of grit of the stone as it catches on the steel causes the chisel to hug the stone and bite in deeper and deeper; while the chisel shown at *N* is thrust away by the action of the grit. Now, friction of any kind is only a sort of grinding operation, and the same principles hold good.

THE NECESSITY FOR GOOD INSTRUMENTS.

It is to be hoped the reader who intends to profit by this treatise has fitted up such a pair of dividers as those we have described, because it is only with accurate instruments he can hope to produce drawings on which any reliance can be placed. The drawing of a ratchet-tooth lever escapement of eight and one-half degrees pallet action will now be resumed. In the drawing at Fig. 18 is shown a complete delineation of such an escapement with eight and one-half degrees of pallet action and equidistant locking faces. It is, of course, understood the escape wheel is to be drawn ten inches in diameter, and that the degree arcs shown in Fig. 1 will be used.

We commence by carefully placing on the drawing-board a sheet of paper about fifteen inches square, and then vertically through the center draw the line a' a''. At some convenient position on this line is established the point a, which represents the center of the escape wheel. In this drawing it is not important that the entire escape wheel be shown, inasmuch as we have really to do with but a little over sixty degrees of the periphery of the escape wheel. With the dividers carefully set at five inches, from a, as a center, we sweep the arc n n, and from the intersection of the perpendicular line a' a'' with the arc n we lay off on each side thirty degrees from the brass degree arc, and through the points thus established are drawn the radial lines a b' and a d'.

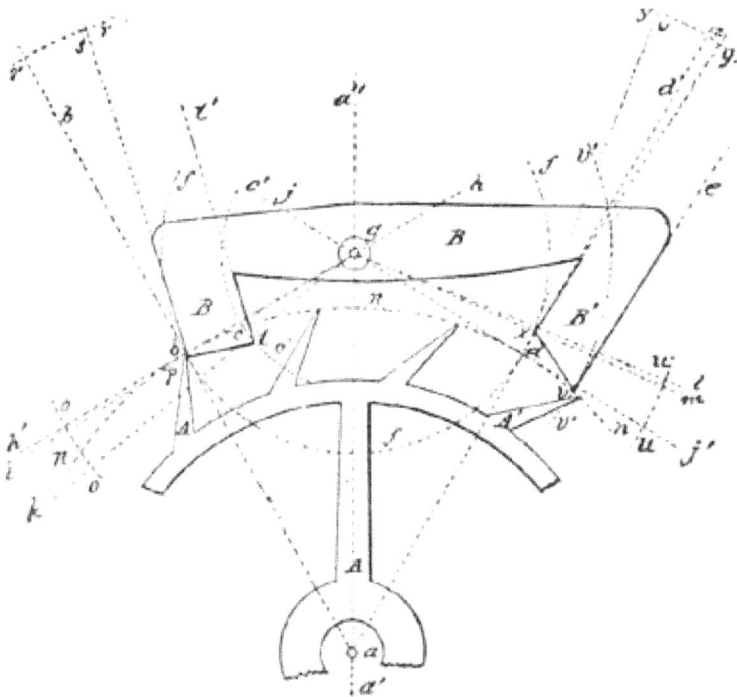

Fig. 18

21

The point on the arc n where it intersects with the line b' is termed the point b. At the intersection of the radial line $a\, d'$ is established the point d. We take ten and one-half degrees in the dividers, and from the point b establish the point c, which embraces the arc of the escape wheel which is utilized by the pallet action. Through the point b the line $h'\, h$ is drawn at right angles to the line $a\, b'$. The line $j\, j'$ is also drawn at right angles to the line $a\, d'$ through the point d. We now have an intersection of the lines just drawn in common with the line $a\, a'$ at the point g, said point indicating the center of the pallet action.

The dividers are now set to embrace the space between the points b and g on the line $h'\, h$, and the arc $f\, f$ is swept; which, in proof of the accuracy of the work, intersects the arc n at the point d. This arc coincides with the locking faces of both pallets. To lay out the entrance pallet, the dividers are set to five inches, and from g as a center the short arc $o\, o$ is swept. On this arc one degree is laid off below the line $h'\, h$, and the line $g\, i$ drawn. The space embraced between the lines h and i on the arc f represents the locking face of the entrance pallet, and the point formed at the intersection of the line $g\, i$ with the arc f is called the point p. To give the proper lock to the face of the pallet, from the point p as a center is swept the short arc $r\, r$, and from its intersection with the line $a\, b'$ twelve degrees are laid off and the line $b\, s$ drawn, which defines the locking face of the entrance pallet. From g as a center is swept the arc $c'\, c'$, intersecting the arc $n\, n$ at c. On this arc (c) is located the inner angle of the entrance pallet. The dividers are set to embrace the space on the arc c' between the lines $g\, h'$ and $g\, k$. With this space in the dividers one leg is set at the point c, measuring down on the arc c' and establishing the point t. The points p and t are then connected, and thus the impulse face of the entrance pallet B is defined. From the point t is drawn the line $t\, t'$, parallel to the line $b\, s$, thus defining the inner face of the entrance pallet.

DELINEATING THE EXIT PALLET.

To delineate the exit pallet, sweep the short arc $u\, u$ (from g as a center) with the dividers set at five inches, and from the intersection of this arc with the line $g\, j'$ set off eight and one-half degrees and draw the line $g\, l$. At one degree below this line is drawn the line $g\, m$. The space on the arc f between these lines defines the locking face of the exit pallet. The point where the line *g m* intersects the arc *f* is named the point *x*. From the point *x* is erected the line *x w*, perpendicular to the line *g m*. From *x* as a center, and with the dividers set at five inches, the short arc *y y* is swept, and on this arc are laid off twelve degrees, and the line *x z* is drawn, which line defines the locking face of the exit pallet.

Next is taken ten and one-half degrees from the brass degree-scale, and from the point *d* on the arc *n* the space named is laid off, and thus is established the point *v*; and from *g* as a center is swept the arc *v' v'* through the point *v*. It will be evident on a little thought, that if the tooth *A'* impelled the exit pallet to the position shown, the outer angle of the pallet must extend down to the point *v*, on the arc *v' v'*; consequently, we define the impulse face of this pallet by drawing a line from point *x* to *v*. To define the outer face of the exit pallet, we draw the line *v e* parallel to the line *x z*.

There are no set rules for drawing the general form of the pallet arms, only to be governed by and conforming to about what we would deem appropriate, and to accord with a sense of proportion and mechanical elegance. Ratchet-tooth pallets are usually made in what is termed "close pallets"; that is, the pallet jewel is set in a slot sawed in the steel pallet arm, which is undoubtedly the strongest and most serviceable form of pallet made. We shall next consider the ratchet-tooth lever escapement with circular pallets and ten degrees of pallet action.

DELINEATING CIRCULAR PALLETS.

To delineate "circular pallets" for a ratchet-tooth lever escapement, we proceed very much as in the former drawing, by locating the point *A*, which represents the center of the escape wheel, at some convenient point, and with the dividers set at five inches, sweep the arc *m*, to represent the periphery of the escape wheel, and then draw the vertical line *A B'*, Fig. 19. We (as before) lay off thirty degrees on the arc *m* each side of the intersection of said arc with the line *A B'*, and thus establish on the arc *m* the points *a b*, and from *A* as a center draw through the points so established the radial lines *A a'* and *A b'*.

We erect from the point *a* a perpendicular to the line *A a*, and, as previously explained, establish the pallet center at *B*. Inasmuch as we are to employ circular pallets, we lay off to the left on the arc *m*, from the point *a*, five degrees, said five degrees being half of the angular motion of the escape wheel utilized in the present drawing, and thus establish the point *c*, and from *A* as a center draw through this point the radial line *A c'*. To the right of the point *a* we lay off five degrees and establish the point *d*. To illustrate the underlying principle of our circular pallets: with one leg of the dividers set at *B* we sweep through the points *c a d* the arcs *c'' a'' d''*.

From *B* as a center, we continue the line *B a* to *f*, and with the dividers set at five inches, sweep the short arc *e e*. From the intersection of this arc with the line *B f* we lay off one and a half degrees and draw the line *B g*, which establishes the extent of the lock on the entrance pallet. It will be noticed the linear extent of the locking face of the entrance pallet is greater than that of the exit, although both represent an angle of one and a half degrees. Really, in practice, this discrepancy is of little importance, as the same side-shake in banking would secure safety in either case.

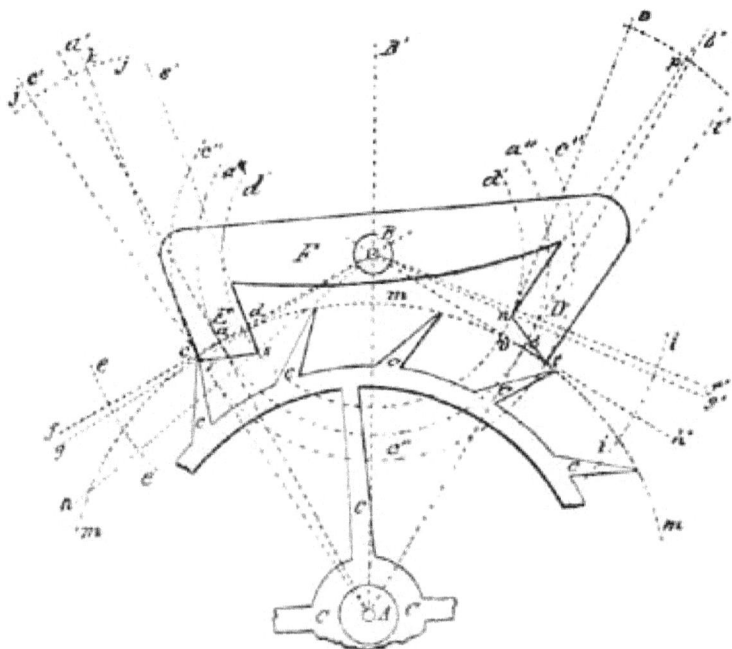

Fig. 19

24

The fault we previously pointed out, of the generally accepted method of delineating a detached lever escapement, is not as conspicuous here as it is where the pallets are drawn with equidistant locking faces; that is, the inner angle of the entrance pallet (shown at s) does not have to be carried down on the arc d' as far to insure a continuous pallet action of ten degrees, as with the pallets with equidistant locking faces. Still, even here we have carried the angle s down about half a degree on the arc d', to secure a safe lock on the exit pallet.

THE AMOUNT OF LOCK.

If we study the large drawing, where we delineate the escape wheel ten inches in diameter, it will readily be seen that although we claim one and a half degrees lock, we really have only about one degree, inasmuch as the curve of the peripheral line m diverges from the line $B f$, and, as a consequence, the absolute lock of the tooth C on the locking face of the entrance pallet E is but about one degree. Under these conditions, if we did not extend the outer angle of the exit pallet at t down to the peripheral line m, we would scarcely secure one-half a degree of lock. This is true of both pallets. We must carry the pallet angles at $r \, s \, n \, t$ down on the circles $c'' \, d'$ if we would secure the lock and impulse we claim; that is, one and a half degrees lock and eight and a half degrees impulse.

Now, while the writer is willing to admit that a one-degree lock in a sound, well-made escapement is ample, still he is not willing to allow of a looseness of drawing to incorporate to the extent of one degree in any mechanical matter demanding such extreme accuracy as the parts of a watch. It has been claimed that such defects can, to a great extent, be remedied by setting the escapement closer; that is, by bringing the centers of the pallet staff and escape wheel nearer together. We hold that such a course is not mechanical and, further, that there is not the slightest necessity for such a policy.

ADVANTAGE OF MAKING LARGE DRAWINGS.

By making the drawings large, as we have already suggested and insisted upon, we can secure an accuracy closely approximating perfection. As, for instance, if we wish to get a lock of one and a half degrees on the locking face of the entrance pallet E, we measure down on the arc c'' from its intersection with the peripheral line m one and a half degrees, and establish the point r and thus locate the outer angle of the entrance pallet E, so there will really be one and a half degrees of lock; and by measuring down on the arc d' ten degrees from its intersection with the peripheral line m, we locate the point s, which determines the position of the inner angle of the entrance pallet, and we know for a certainty that when this inner angle is freed from the tooth it will be after the pallet (and, of course, the lever) has passed through exactly ten degrees of angular motion.

For locating the inner angle of the exit pallet, we measure on the arc d', from its intersection with the peripheral line m, eight and a half degrees, and establish the point n, which locates the position of this inner angle; and, of course, one and a half degrees added on the arc d' indicates the extent of the lock on this pallet. Such drawings not only enable us to theorize to extreme exactness, but also give us proportionate measurements, which can be carried into actual construction.

THE CLUB-TOOTH LEVER ESCAPEMENT.

We will now take up the club-tooth form of the lever escapement. This form of tooth has in the United States and in Switzerland almost entirely superceded the ratchet tooth. The principal reason for its finding so much favor is, we think, chiefly owing to the fact that this form of tooth is better able to stand the manipulations of the able-bodied watchmaker, who possesses more strength than skill. We will not pause now, however, to consider the comparative merits of the ratchet and club-tooth forms of the lever escapement, but leave this part of the theme for discussion after we have given full instructions for delineating both forms.

With the ratchet-tooth lever escapement all of the impulse must be derived from the pallets, but in the club-tooth escapement we can divide the impulse planes between the pallets and the teeth to suit our fancy; or perhaps it would be better to say carry out theories, because we have it in our power, in this form of the lever escapement, to indulge ourselves in many changes of the relations of the several parts. With the ratchet tooth the principal changes we could make would be from pallets with equidistant lockings to circular pallets. The club-

26

tooth escape wheel not only allows of circular pallets and equidistant lockings, but we can divide the impulse between the pallets and the teeth in such a way as will carry out many theoretical advantages which, after a full knowledge of the escapement action is acquired, will naturally suggest themselves. In the escapement shown at Fig. 20 we have selected, as a very excellent example of this form of tooth, circular pallets of ten degrees fork action and ten and a half degrees of escape-wheel action.

It will be noticed that the pallets here are comparatively thin to those in general use; this condition is accomplished by deriving the principal part of the impulse from driving planes placed on the teeth. As relates to the escape-wheel action of the ten and one-half degrees, which gives impulse to the escapement, five and one-half degrees are utilized by the driving planes on the teeth and five by the impulse face of the pallet. Of the ten degrees of fork action, four and a half degrees relate to the impulse face of the teeth, one and a half degrees to lock, and four degrees to the driving plane of the pallets.

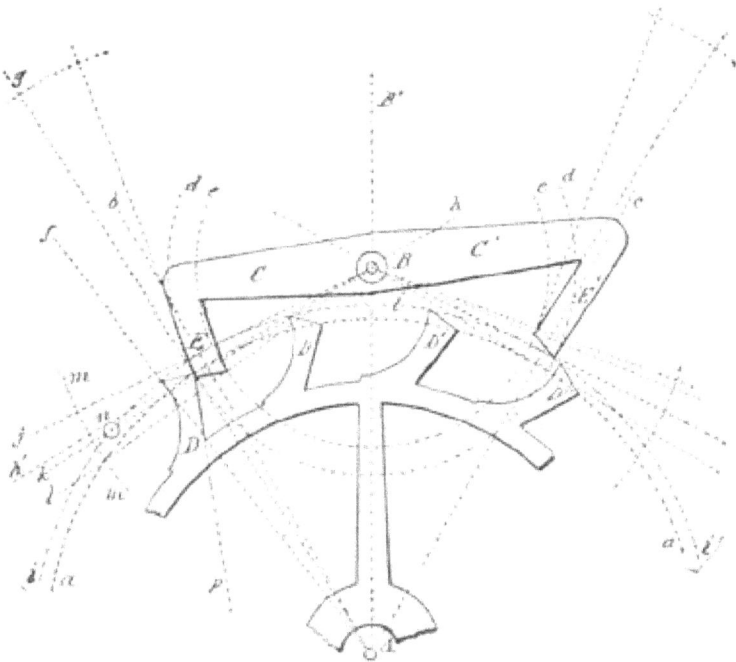

Fig. 20

27

In delineating such a club-tooth escapement, we commence, as in former examples, by first assuming the center of the escape wheel at *A*, and with the dividers set at five inches sweeping the arc *a a*. Through *A* we draw the vertical line *A B'*. On the arc *a a*, and each side of its intersection with the line *A B'*, we lay off thirty degrees, as in former drawings, and through the points so established on the arc *a a* we draw the radial lines *A b* and *A c*. From the intersection of the radial line *A b* with the arc *a* we draw the line *h h* at right angles to *A b*. Where the line *h* intersects the radial lines *A B'* is located the center of the pallet staff, as shown at *B*. Inasmuch as we decided to let the pallet utilize five degrees of escape-wheel action, we take a space of two and a half degrees in the dividers, and on the arc *a a* lay off the said two and a half degrees to the left of this intersection, and through the point so established draw the radial line *A g*. From *B* as a center we sweep the arc *d d* so it passes through the point of intersection of the arc *a* with the line *A g*.

We again lay off two and a half degrees from the intersection of the line *A b* with the arc *a*, but this time to the right of said intersection, and through the point so established, and from *B* as a center, we sweep the arc *e*. From the intersection of the radial line *A g* with the arc *a* we lay off to the left five and a half degrees on said arc, and through the point so established draw the radial line *A f*. With the dividers set at five inches we sweep the short arc *m* from *B* as a center. From the intersection of the line *h B h'* with the arc *m* we lay off on said arc and above the line *h'* four and a half degrees, and through the point so established draw the line *B j*.

We next set the dividers so they embrace the space on the radial line *A b* between its intersection with the line *B j* and the center *A*, and from *A* as a center sweep the arc *i*, said arc defining the *addendum* of the escape-wheel teeth. We draw a line from the intersection of the radial line *A f* with the arc *i* to the intersection of the radial line *A g* with the arc *a*, and thus define the impulse face of the escape-wheel tooth *D*. For defining the locking face of the tooth we draw a line at an angle of twenty-four degrees to the line *A g*, as previously described. The back of the tooth is defined with a curve swept from some point on the addendum circle *i*, such as our judgment will dictate.

28

In the drawing shown at Fig. 20 the radius of this curve was obtained by taking eleven and a half degrees from the degree arc of 5" radius in the dividers, and setting one leg at the intersection of the radial line $A\,f$ with the arc i, and placing the other on the line i, and allowing the point so established to serve as a center, the arc was swept for the back of the tooth, the small circle at n denoting one of the centers just described. The length for the face of the tooth was obtained by taking eleven degrees from the degree arc just referred to and laying that space off on the line p, which defined the face of the tooth. The line $B\,k$ is laid off one and a half degrees below $B\,h$ on the arc m. The extent of this arc on the arc d defines the locking face of the entrance pallet. We set off four degrees on the arc m below the line $B\,k$, and through the point so established draw the line $B\,l$. We draw a line from the intersection of the line $A\,g$ with the line $c\,h$ to the intersection of the arc e with the line $c\,l$, and define the impulse face of the entrance pallet.

RELATIONS OF THE SEVERAL PARTS.

Before we proceed to delineate the exit pallet of our escapement, let us reason on the relations of the several parts.

The club-tooth lever escapement is really the most complicated escapement made. We mean by this that there are more factors involved in the problem of designing it correctly than in any other known escapement. Most—we had better say all, for there are no exceptions which occur to us—writers on the lever escapement lay down certain empirical rules for delineating the several parts, without giving reasons for this or that course. For illustration, it is an established practice among escapement makers to employ tangential lockings, as we explained and illustrated in Fig. 16.

Now, when we adopt circular pallets and carry the locking face of the entrance pallet around to the left two and a half degrees, the true center for the pallet staff, if we employ tangent lockings, would be located on a line drawn tangent to the circle $a\,a$ from its intersection with the radial line $A\,k$, Fig. 21. Such a tangent is depicted at the line $s\,l'$. If we reason on the situation, we will see that the line $A\,k$ is not at right angles to the line $s\,l$; and, consequently, the locking face of the entrance pallet *E* has not really the twelve-degree lock we are taught to believe it has.

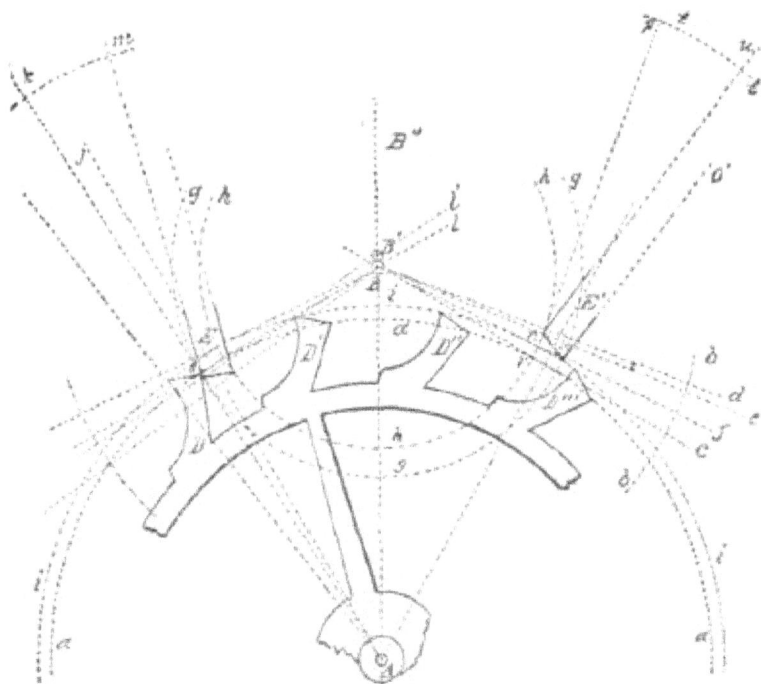

Fig. 21

We will not discuss these minor points further at present, but leave them for subsequent consideration. We will say, however, that we could locate the center of the pallet action at the small circle *B'* above the center *B*, which we have selected as our fork-and-pallet action, and secure a perfectly sound escapement, with several claimed advantages.

Let us now take up the delineation of the exit pallet. It is very easy to locate the outer angle of this pallet, as this must be situated at the intersection of the addendum circle *i* and the arc *g*, and located at *o*. It is also self-evident that the inner or locking angle must be situated at some point on the arc *h*. To determine this location we draw the line *B c* from *B* (the pallet center) through the intersection of the arc *h* with the pitch circle *a*.

30

Again, it follows as a self-evident fact, if the pallet we are dealing with was locked, that is, engaged with the tooth D'', the inner angle n of the exit pallet would be one and a half degrees inside the pitch circle a. With the dividers set at 5", we sweep the short arc b b, and from the intersection of this arc with the line B c we lay off ten degrees, and through the point so established, from B, we draw the line B d. Below the point of intersection of the line B d with the short arc b b we lay off one and a half degrees, and through the point thus established we draw the line B e.

LOCATING THE INNER ANGLE OF THE EXIT PALLET.

The intersection of the line B e with the arc h, which we will term the point n, represents the location of the inner angle of the exit pallet. We have already explained how we located the position of the outer angle at o. We draw the line n o and define the impulse face of the exit pallet. If we mentally analyze the problem in hand, we will see that as the exit pallet vibrates through its ten degrees of arc the line B d and B c change places, and the tooth D'' locks one and a half degrees. To delineate the locking face of the exit pallet, we erect a perpendicular to the line B e from the point n, as shown by the line n p.

From n as a center we sweep the short arc t t, and from its intersection with the line n p we lay off twelve degrees, and through the point so established we draw the line n u, which defines the locking face of the exit pallet. We draw the line o o' parallel with n u and define the outer face of said pallet. In Fig. 21 we have not made any attempt to show the full outline of the pallets, as they are delineated in precisely the same manner as those previously shown.

We shall next describe the delineation of a club-tooth escapement with pallets having equidistant locking faces; and in Fig. 22 we shall show pallets with much wider arms, because, in this instance, we shall derive more of the impulse from the pallets than from the teeth. We do this to show the horological student the facility with which the club-tooth lever escapement can be manipulated. We wish also to impress on his mind the facts that the employment of thick pallet arms and thin pallet arms depends on the teeth of the escape wheel for its efficiency, and that he must have knowledge enough of the principles of action to tell at a glance on what lines the escapement was constructed.

Suppose, for illustration, we get hold of a watch which has thin pallet arms, or stones, if they are exposed pallets, and the escape was designed for pallets with thick arms. There is no sort of tinkering we can do to give such a watch a good motion, except to change either the escape wheel or the pallets. If we know enough of the lever escapement to set about it with skill and judgment, the matter is soon put to rights; but otherwise we can look and squint, open and close the bankings, and tinker about till doomsday, and the watch be none the better.

CLUB-TOOTH LEVER WITH EQUIDISTANT LOCKING FACES.

In drawing a club-tooth lever escapement with equidistant locking, we commence, as on former occasions, by producing the vertical line *A k*, Fig. 22, and establishing the center of the escape wheel at *A*, and with the dividers set at 5" sweep the pitch circle *a*. On each side of the intersection of the vertical line *A k* with the arc *a* we set off thirty degrees on said arc, and through the points so established draw the radial lines *A b* and *A c*.

From the intersection of the radial line *A b* with the arc *a* lay off three and a half degrees to the left of said intersection on the arc *a*, and through the point so established draw the radial line *A e*. From the intersection of the radial line *A b* with the arc *a* erect the perpendicular line *f*, and at the crossing or intersection of said line with the vertical line *A k* establish the center of the pallet staff, as indicated by the small circle *B*. From *B* as a center sweep the short arc *l* with a 5" radius; and from the intersection of the radial line *A b* with the arc *a* continue the line *f* until it crosses the short arc *l*, as shown at *f'*. Lay off one and a half degrees on the arc *l* below its intersection with the line *f'*, and from *B* as a center draw the line *B i* through said intersection. From *B* as a center, through the intersection of the radial line *A b* and the arc *a*, sweep the arc *g*.

The space between the lines *B f'* and *B i* on the arc *g* defines the extent of the locking face of the entrance pallet *C*. The intersection

32

of the line $B f'$ with the arc g we denominate the point o, and from this point as a center sweep the short arc p with a 5" radius; and on this arc, from its intersection with the radial line A b, lay off twelve degrees, and through the point so established, from o as a center, draw the radial line o m, said line defining the locking face of the entrance pallet C.

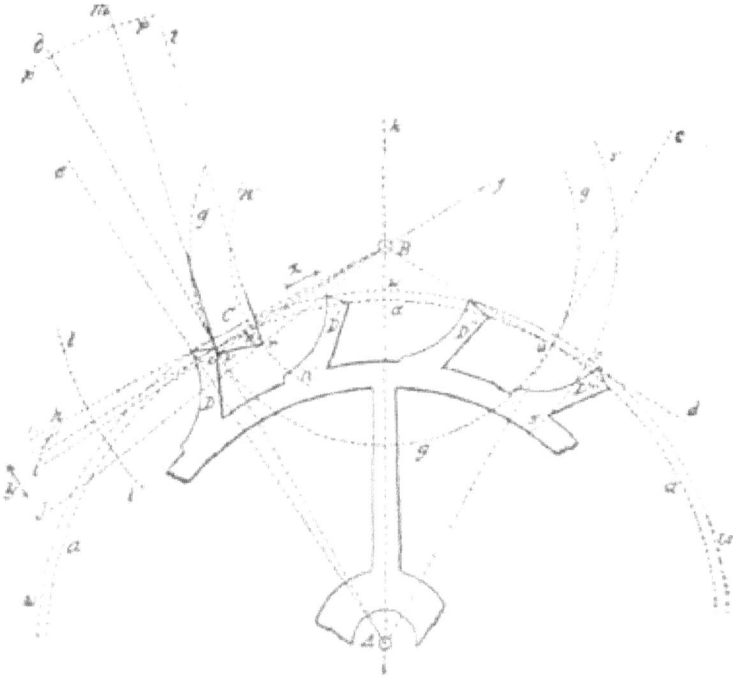

Fig. 22

It will be seen that this gives a positive "draw" of twelve degrees to the entrance pallet; that is, counting to the line B f'. In this escapement as delineated there is perfect tangential locking. If the locking face of the entrance-pallet stone at C was made to conform to the radial line A b, the lock of the tooth D at o would be "dead"; that is, absolutely neutral. The tooth D would press the pallet C in the direction of the arrow x, toward the center of the pallet staff B, with no tendency on the part of the pallet to turn on its axis B. Theoretically, the pallet with the locking face cut to coincide with the line A b would resist movement on the center B in either direction indicated by the double-headed arrow y.

A pallet at C with a circular locking face made to conform to the arc g, would permit movement in the direction of the double-headed arrow y with only mechanical effort enough to overcome friction. But it is evident on inspection that a locking face on the line $A\ b$ would cause a retrograde motion of the escape wheel, and consequent resistance, if said pallet was moved in either direction indicated by the double-headed arrow y. Precisely the same conditions obtain at the point u, which holds the same relations to the exit pallet as the point o does to the entrance pallet C.

ANGULAR MOTION OF ESCAPE WHEEL DETERMINED.

The arc (three and a half degrees) of the circle a embraced between the radial lines $A\ b$ and $A\ e$ determines the angular motion of the escape wheel utilized by the escape-wheel tooth. To establish and define the extent of angular motion of the escape wheel utilized by the pallet, we lay off seven degrees on the arc a from the point o and establish the point n, and through the point n, from B as a center, we sweep the short arc n'. Now somewhere on this arc n' will be located the inner angle of the entrance pallet. With a carefully-made drawing, having the escape wheel 10" in diameter, it will be seen that the arc a separates considerably from the line, $B\ f'$ where it crosses the arc n'.

It will be remembered that when drawing the ratchet-tooth lever escapement a measurement of eight and a half degrees was made on the arc n' down from its intersection with the pitch circle, and thus the inner angle of the pallet was located. In the present instance the addendum line w becomes the controlling arc, and it will be further noticed on the large drawing that the line $B\ h$ at its intersection with the arc n' approaches nearer to the arc w than does the line $B\ f'$ to the pitch circle a; consequently, the inner angle of the pallet should not in this instance be carried down on the arc n' so far to correct the error as in the ratchet tooth.

Reason tells us that if we measure ten degrees down on the arc n' from its intersection with the addendum circle w we must define the

34

position of the inner angle of the entrance pallet. We name the point so established the point *r*. The outer angle of this pallet is located at the intersection of the radial line *A b* with the line *B i*; said intersection we name the point *v*. Draw a line from the point *v* to the point *r*, and we define the impulse face of the entrance pallet; and the angular motion obtained from it as relates to the pallet staff embraces six degrees.

Measured on the arc *l*, the entire ten degrees of angular motion is as follows: Two and a half degrees from the impulse face of the tooth, and indicated between the lines *B h* and *B f*; one and a half degrees lock between the lines *B f'* and *B i*; six degrees impulse from pallet face, entrance between the lines *B i* and *B j*.

A DEPARTURE FROM FORMER PRACTICES.

Grossmann and Britten, in all their delineations of the club-tooth escapement, show the exit pallet as disengaged. To vary from this beaten track we will draw our exit pallet as locked. There are other reasons which prompt us to do this, one of which is, pupils are apt to fall into a rut and only learn to do things a certain way, and that way just as they are instructed.

To illustrate, the writer has met several students of the lever escapement who could make drawings of either club or ratchet-tooth escapement with the lock on the entrance pallet; but when required to draw a pallet as illustrated at Fig. 23, could not do it correctly. Occasionally one could do it, but the instances were rare. A still greater poser was to request them to delineate a pallet and tooth when the action of escaping was one-half or one-third performed; and it is easy to understand that only by such studies the master workman can thoroughly comprehend the complications involved in the club-tooth lever escapement.

AN APT ILLUSTRATION.

As an illustration: Two draughtsmen, employed by two competing watch factories, each designs a club-tooth escapement. We will further suppose the trains and mainspring power used by each concern to be precisely alike. But in practice the escapement of the watches made by one factory would "set," that is, if you stopped the balance dead still, with the pin in the fork, the watch would not start of itself; while the escapement designed by the other draughtsman would not "set"—stop the balance dead as often as you choose, the watch would start of itself. Yet even to experienced workmen the escape wheels and pallets *looked* exactly alike. Of course, there was a difference, and still none of the text-books make mention of it.

For the present we will go on with delineating our exit pallet. The preliminaries are the same as with former drawings, the instructions for which we need not repeat. Previous to drawing the exit pallet, let us reason on the matter. The point *r* in Fig. 23 is located at the intersection of pitch circle *a* and the radial line *A c*; and this will also be the point at which the tooth *C* will engage the locking face of the exit pallet.

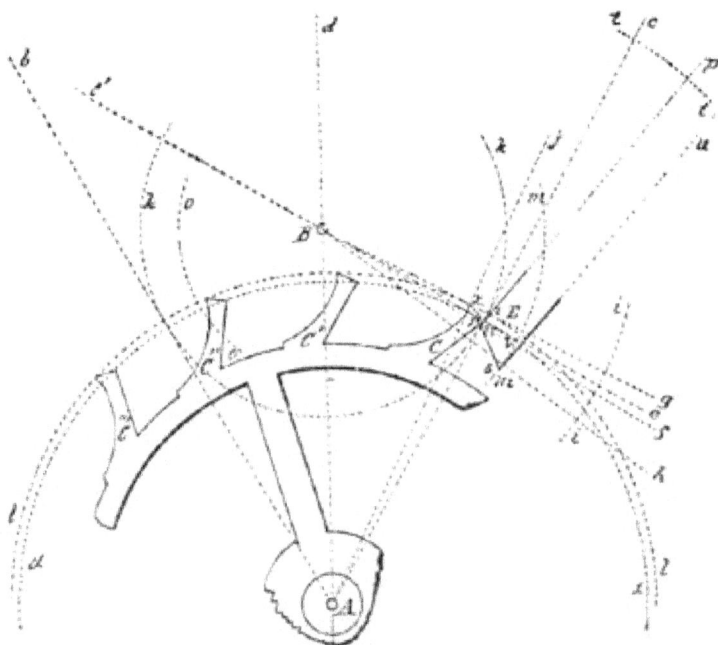

Fig. 23

This point likewise represents the advance angle of the engaging tooth. Now if we measure on the arc *k* (which represents the locking faces of both pallets) downward one and a half degrees, we establish the lock of the pallet *E*. To get this one and a half degrees

36

defined on the arc k, we set the dividers at 5", and from B as a center sweep the short arc i, and from the intersection of the arc i with the line B e we lay off on said arc i one and a half degrees, and through the point so established draw the line $B f$.

Now the space on the arc k between the lines B e and $B f$ defines the angular extent of the locking face. With the dividers set at 5" and one leg resting at the point r, we sweep the short arc t, and from the intersection of said arc with the line A c we draw the line n p; but in doing so we extend it (the line) so that it intersects the line $B f$, and at said intersection is located the inner angle of the exit pallet. This intersection we will name the point n.

From the intersection of the line B e with the arc i we lay off two and a half degrees on said arc, and through the point so established we draw the line B g. The intersection of this line with the arc k we name the point z. With one leg of our dividers set at A we sweep the arc l so it passes through the point z. This last arc defines the addendum of the escape-wheel teeth. From the point r on the arc a we lay off three and a half degrees, and through the point so established draw the line $A j$.

LOCATING THE OUTER ANGLE OF THE IMPULSE PLANES.

The intersection of this line with the addendum arc l locates the outer angle of the impulse planes of the teeth, and we name it the point x. From the point r we lay off on the arc a seven degrees and establish the point v, which defines the extent of the angular motion of the escape wheel utilized by pallet. Through the point v, from B as a center, we sweep the short arc m. It will be evident on a moment's reflection that this arc m must represent the path of movement of the outer angle of the exit pallet, and if we measure down ten degrees from the intersection of the arc l with the arc m, the point so established (which we name the point s) must be the exact position of the outer angle of the pallet during locking. We have a measure of ten degrees on the arc m, between the lines B g and B h, and by taking this space in the dividers and setting one leg at the intersection of the arc l with the arc m, and measuring down on m, we establish the point s.

Drawing a line from point *n* to point *s* we define the impulse face of the pallet.

MAKING AN ESCAPEMENT MODEL.

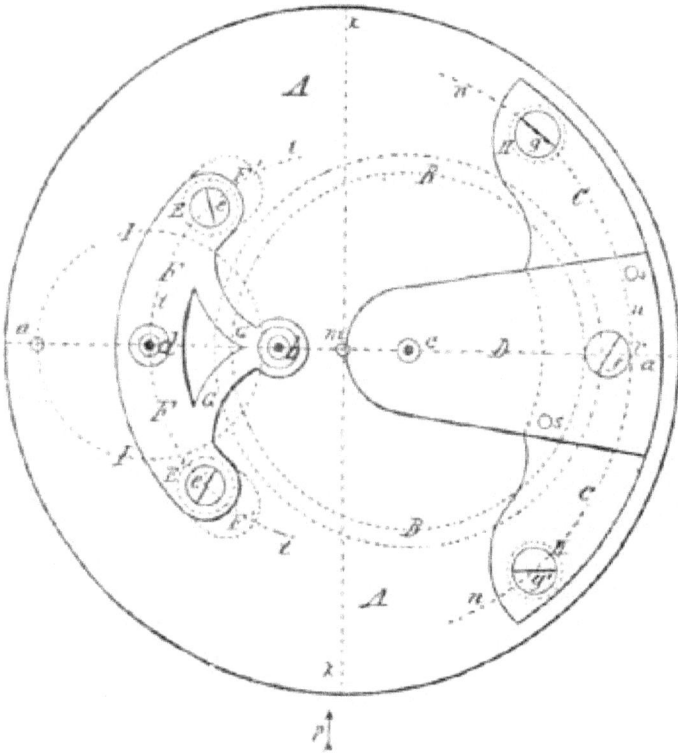

Fig. 24

It is next proposed we apply the theories we have been considering and make an enlarged model of an escapement, as shown at Figs. 24 and 25. This model is supposed to have an escape wheel one-fifth the size of the 10" one we have been drawing. In the accompanying cuts are shown only the main plate and bridges in full lines, while the positions of the escape wheel and balance are indicated by the dotted circles *I B*. The cuts are to no precise scale, but were reduced from a full-size drawing for convenience in printing. We shall give exact dimensions, however, so there will be no difficulty in carrying out our instructions in construction.

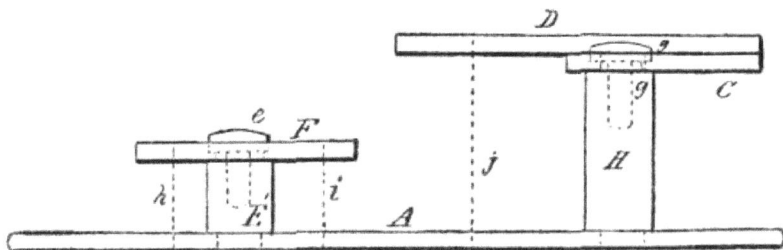

Fig. 25

Perhaps it would be as well to give a general description of the model before taking up the details. A reduced side view of the complete model is given at Fig. 26. In this cut the escapement model shown at Figs. 24 and 25 is sketched in a rough way at *R*, while *N* shows a glass cover, and *M* a wooden base of polished oak or walnut. This base is recessed on the lower side to receive an eight-day spring clock movement, which supplies the motive power for the model. This base is recessed on top to receive the main plate *A*, Fig. 24, and also to hold the glass shade *N* in position. The base *M* is 2½" high and 8" diameter. The glass cover *N* can have either a high and spherical top, as shown, or, as most people prefer, a flattened oval.

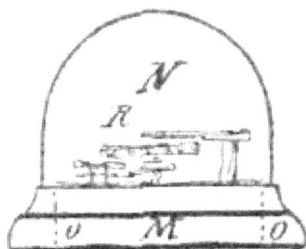

Fig. 26

The main plate *A* is of hard spring brass, 1/10" thick and 6" in diameter; in fact, a simple disk of the size named, with slightly rounded edges. The top plate, shown at *C*, Figs. 24 and 25, is 1/8" thick and shaped as shown. This plate (*C*) is supported on two pillars 1/2" in diameter and 1-1/4" high. Fig. 25 is a side view of Fig. 24 seen in the direction of the arrow *p*. The cock *D* is also of 1/8" spring brass shaped as shown, and attached by the screw *f* and steady pins *s s* to the top plate *C*. The bridge *F G* carries the top pivots of escape wheel and pallet staff, and is shaped as shown at the full outline. This bridge is supported on two pillars 1/2" high and 1/2" in diameter, one of which is shown at *E*, Fig. 25, and both at the dotted circles *E E'*, Fig. 24.

To lay out the lower plate we draw the line *a a* so it passes through the center of *A* at *m*. At 1.3" from one edge of *A* we establish on the line *a* the point *d*, which locates the center of the escape wheel. On the same line *a* at 1.15" from *d* we establish the point *b*, which represents the center of the pallet staff. At the distance of 1.16" from *b* we establish the point *c*, which represents the center of the balance staff. To locate the pillars *H*, which support the top plate *C*, we set the dividers at 2.58", and from the center *m* sweep the arc *n*.

From the intersection of this arc with the line *a* (at *r*) we lay off on said arc *n* 2.1" and establish the points *g g'*, which locate the center of the pillars *H H*. With the dividers set so one leg rests at the center *m* and the other leg at the point *d*, we sweep the arc *t*. With the dividers set at 1.33" we establish on the arc *t*, from the point *d*, the points *e e'*, which locate the position of the pillars *E E'*. The outside diameter of the balance *B* is 3-5/8" with the rim 3/16" wide and 5/16" deep, with screws in the rim in imitation of the ordinary compensation balance.

Speaking of a balance of this kind suggests to the writer the trouble he experienced in procuring material for a model of this kind— for the balance, a pattern had to be made, then a casting made, then a machinist turned the casting up, as it was too large for an American lathe. A hairspring had to be specially made, inasmuch as a mainspring was too short, the coils too open and, more particularly, did not look well. Pallet jewels had to be made, and lapidists have usually poor ideas of close measurements. Present-day conditions, however, will, no doubt, enable the workman to follow our instructions much more readily.

MAKING THE BRIDGES.

In case the reader makes the bridges *C* and *F*, as shown in Fig. 27, he should locate small circles on them to indicate the position of the screws for securing these bridges to the pillars which support them, and also other small circles to indicate the position of the pivot holes d b for the escape wheel and pallet staff. In practice it will be well to draw the line a a through the center of the main plate A, as previously directed, and also establish the point d as therein directed.

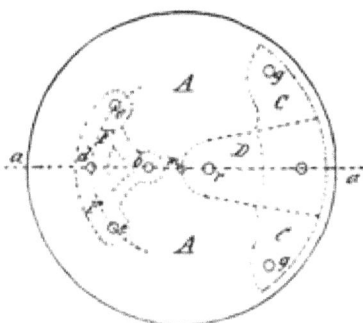

Fig. 27

The pivot hole d' for the escape wheel, and also the holes at e e and b, are now drilled in the bridge F. These holes should be about 1/16" in diameter. The same sized hole is also drilled in the main plate A at d. We now place a nicely-fitting steel pin in the hole d' in the bridge F and let it extend into the hole d in the main plate. We clamp the bridge F to A so the hole b comes central on the line a, and using the holes e e in F as guides, drill or mark the corresponding holes e' e' and b in the main plate for the pillars E E' and the pallet staff.

This plan will insure the escape wheel and pallet staff being perfectly upright. The same course pursued with the plate C will insure the balance being upright. The pillars which support the bridges are shaped as shown at Fig. 28, which shows a side view of one of the pillars which support the top plate or bridge C. The ends are turned to 1/4" in diameter and extend half through the plate, where they are held by screws, the same as in American movements.

The pillars (like H) can be riveted in the lower plate A, but we think most workmen will find it more satisfactory to employ screws, as shown at Fig. 29. The heads of such screws should be about 3/8" in diameter and nicely rounded, polished and blued. We would not advise jeweling the pivot holes, because there is but slight friction, except to

41

the foot of the balance pivot, which should be jeweled with a plano-convex garnet.

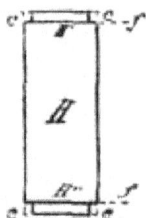

Fig. 28

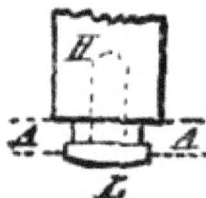

Fig. 29

IMITATION RUBIES FOR CAPPING THE TOP PIVOTS.

The top pivots to the escape wheel should be capped with imitation rubies for appearance sake only, letting the cap settings be red gold, or brass red gilded. If real twelve-karat gold is employed the cost will not be much, as the settings are only about 3/8" across and can be turned very thin, so they will really contain but very little gold. The reason why we recommend imitation ruby cap jewels for the upper holes, is that such jewels are much more brilliant than any real stone we can get for a moderate cost. Besides, there is no wear on them.

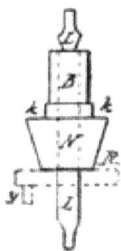

Fig. 30

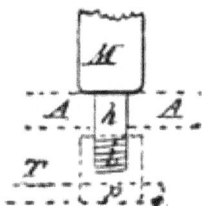

Fig. 31

The pallet jewels are also best made of glass, as garnet or any red stone will look almost black in such large pieces. Red carnelian has a sort of brick-red color, which has a cheap appearance. There is a new phosphorus glass used by optical instrument makers which is intensely hard, and if colored ruby-red makes a beautiful pallet jewel, which will afford as much service as if real stones were used; they are no cheaper than carnelian pallets, but much richer looking. The prettiest cap for the balance is one of those foilback stones in imitation of a rose-cut diamond.

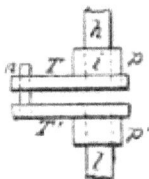

Fig. 32

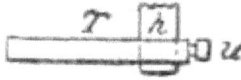

Fig. 33

In turning the staffs it is the best plan to use double centers, but a piece of Stubs steel wire that will go into a No. 40 wire chuck, will answer; in case such wire is used, a brass collet must be provided. This will be understood by inspecting Fig. 30, where L represents the Stubs wire and $B\ N$ the brass collet, with the balance seat shown at k. The escape-wheel arbor and pallet staff can be made in the same way. The lower end of the escape wheel pivot is made about 1/4" long, so that a short piece of brass wire can be screwed upon it, as shown in Fig. 31, where h represents the pivot, A the lower plate, and the dotted line at p the brass piece screwed on the end of the pivot. This piece p is simply a short bit of brass wire with a female screw tapped into the end, which screws on to the pivot. An arm is attached to p, as shown at T. The idea is, the pieces $T\ p$ act like a lathe dog to convey the power from one of the pivots of an old eight-day spring clock movement, which is secured by screws to the lower side of the main plate A. The plan is illustrated at Fig. 32, where l represents pivot of the eight-day clock employed to run the model. Counting the escape-wheel pivot of the clock as one, we take the third pivot from this in the clock train, placing the movement so this point comes opposite the escape-wheel pivot of the model, and screw the clock movement fast to the lower side of the plate A. The parts T, Fig. 33, are alike on both pivots.

PROFITABLE FOR EXPLAINING TO A CUSTOMER.

To fully appreciate such a large escapement model as we have been describing, a person must see it with its great balance, nearly 4" across, flashing and sparkling in the show window in the evening, and the brilliant imitation ruby pallets dipping in and out of the escape wheel. A model of this kind is far more attractive than if the entire train were shown, the mystery of "What makes it go?" being one of the attractions. Such a model is, further, of great value in explaining to a customer what you mean when you say the escapement of his watch is out of order. Any practical workman can easily make an even $100 extra in a year by making use of such a model.

For explaining to customers an extra balance cock can be used to show how the jewels (hole and cap) are arranged. Where the parts are as large as they are in the model, the customer can see and understand for himself what is necessary to be done.

It is not to be understood that our advice to purchase the jewels for an extra balance cock conflicts with our recommending the reader not to jewel the holes of his model. The extra cock is to be shown, not for use, and is employed solely for explaining to a customer what is required when a pivot or jewel is found to be broken.

HOW LARGE SCREWS ARE MADE.

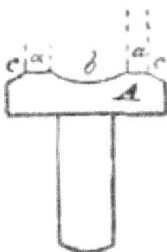

Fig. 34

Fig. 35

The screws which hold the plates in place should have heads about 3/8" in diameter, to be in proportion to the scale on which the balance and escape wheel are gotten up. There is much in the manner in which the screw heads are finished as regards the elegance of such a model. A perfectly flat head, no matter how highly polished, does not look well, neither does a flattened conehead, like Fig. 35. The best head for this purpose is a cupped head with chamfered edges, as shown at Fig. 34 in vertical section. The center *b* is ground and polished into a perfect concave by means of a metal ball. The face, between the lines *a a*, is polished dead flat, and the chamfered edge *a c* finished a trifle convex. The flat surface at *a* is bright, but the concave *b* and chamfer at *c* are beautifully blued. For a gilt-edged, double extra head, the chamfer at *c* can be "snailed," that is, ground with a suitable lap before bluing, like the stem-wind wheels on some watches.

FANCY SCREWHEADS.

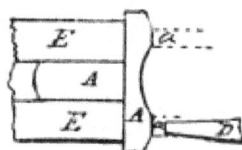

Fig. 36

There are two easy methods of removing the blue from the flat part of the screwhead at *a*. (1) Make a special holder for the screw in the end of a cement brass, as shown at *E*, Fig. 36, and while it is slowly revolving in the lathe touch the flat surface *a* with a sharpened pegwood wet with muriatic acid, which dissolves the blue coating of oxide of iron. (2) The surface of the screwhead is coated with a very thin coating of shellac dissolved in alcohol and thoroughly dried, or a thin coating of collodion, which is also dried. The screw is placed in the ordinary polishing triangle and the flat face at *a* polished on a tin lap with diamantine and oil. In polishing such surfaces the thinnest possible coating of diamantine and oil is smeared on the lap—in fact, only enough to dim the surface of the tin. It is, of course, understood that it is necessary to move only next to nothing of the material to restore the polish of the steel. The polishing of the other steel parts is done precisely like any other steel work.

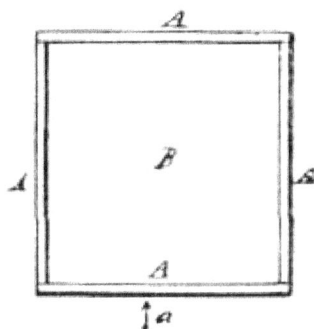

Fig. 37

The regulator is of the Howard pattern. The hairspring stud is set in the cock like the Elgin three-quarter-plate movement. The richest finish for such a model is frosted plates and bridges. The frosting should not be a fine mat, like a watch movement, but coarse-grained—in fact, the grain of the frosting should be proportionate to the size of the movement. The edges of the bridges and balance cock can be left smooth. The best process for frosting is by acid. Details for doing the work will now be given.

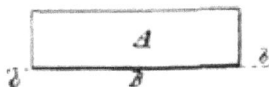

Fig. 38

To do this frosting by acid nicely, make a sieve by tacking and gluing four pieces of thin wood together, to make a rectangular box without a bottom. Four pieces of cigar-box wood, 8" long by 1-1/2" wide, answer first rate. We show at A A A A, Fig. 37, such a box as if seen from above; with a side view, as if seen in the direction of the arrow a, at Fig. 38. A piece of India muslin is glued across the bottom, as shown at the dotted lines b b. By turning up the edges on the outside of the box, the muslin bottom can be drawn as tight as a drum head.

47

HOW TO DO ACID FROSTING.

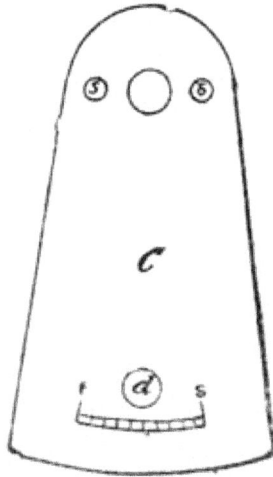

Fig. 39

To do acid frosting, we procure two ounces of gum mastic and place in the square sieve, shown at Fig. 37. Usually more than half the weight of gum mastic is in fine dust, and if not, that is, if the gum is in the shape of small round pellets called "mastic tears," crush these into dust and place the dust in A. Let us next suppose we wish to frost the cock on the balance, shown at Fig. 39. Before we commence to frost, the cock should be perfectly finished, with all the holes made, the regulator cap in position, the screw hole made for the Howard regulator and the index arc engraved with the letters S and F.

It is not necessary the brass should be polished, but every file mark and scratch should be stoned out with a Scotch stone; in fact, be in the condition known as "in the gray." It is not necessary to frost any portion of the cock C, except the upper surface. To protect the portion of the cock not to be frosted, like the edges and the back, we "stop out" by painting over with shellac dissolved in alcohol, to which a little lampblack is added. It is not necessary the coating of shellac should be very thick, but it is important it should be well dried.

48

HOW TO PREPARE THE SURFACE.

For illustration, let us suppose the back and edges of the cock at Fig. 39 are coated with shellac and it is laid flat on a piece of paper about a foot square to catch the excess of mastic. Holes should be made in this paper and also in the board on which the paper rests to receive the steady pins of the cock. We hold the sieve containing the mastic over the cock and, gently tapping the box *A* with a piece of wood like a medium-sized file handle, shake down a little snowstorm of mastic dust over the face of the cock *C*.

Exactly how much mastic dust is required to produce a nice frosting is only to be determined by practice. The way to obtain the knack is to frost a few scraps to "get your hand in." Nitric acid of full strength is used, dipping the piece into a shallow dish for a few seconds. A good-sized soup plate would answer very nicely for frosting the bottom plate, which, it will be remembered, is 6" in diameter.

HOW TO ETCH THE SURFACE.

After the mastic is sifted on, the cock should be heated up to about 250° F., to cause the particles of mastic to adhere to the surface. The philosophy of the process is, the nitric acid eats or dissolves the brass, leaving a little brass island the size of the particle of mastic which was attached to the surface. After heating to attach the particles of mastic, the dipping in nitric acid is done as just described. Common commercial nitric acid is used, it not being necessary to employ chemically pure acid. For that matter, for such purposes the commercial acid is the best.

After the acid has acted for fifteen or twenty seconds the brass is rinsed in pure water to remove the acid, and dried by patting with an old soft towel, and further dried by waving through the air. A little turpentine on a rag will remove the mastic, but turpentine will not touch the shellac coating. The surface of the brass will be found irregularly acted upon, producing a sort of mottled look. To obtain a nice frosting the process of applying the mastic and etching must be repeated three or four times, when a beautiful coarse-grain mat or frosting will be produced.

The shellac protection will not need much patching up during the three or four bitings of acid, as the turpentine used to wash off the mastic does not much affect the shellac coating. All the screw holes like *s s* and *d*, also the steady pins on the back, are protected by varnishing with shellac. The edges of the cocks and bridges should be polished by rubbing lengthwise with willow charcoal or a bit of chamois skin saturated with oil and a little hard rouge scattered upon it. The frosting needs thorough scratch-brushing.

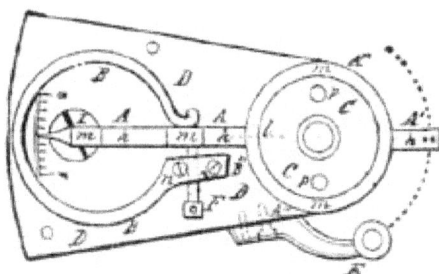

Fig. 40

At Fig. 40 we show the balance cock of our model with modified form of Howard regulator. The regulator bar *A* and spring *B* should be ground smooth on one side and deeply outlined to perfect form. The regulator cap *C* is cut out to the correct size. These parts are of decarbonized cast steel, annealed until almost as soft as sheet brass. It is not so much work to finish these parts as one might imagine. Let us take the regulator bar for an example and carry it through the process of making. The strip of soft sheet steel on which the regulator bar is outlined is represented by the dotted outline *b*, Fig. 41.

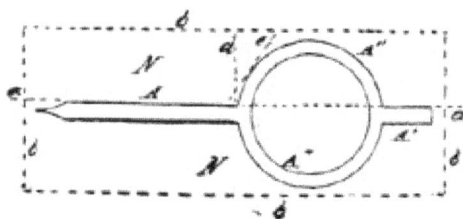

Fig. 41

To cut out sheet steel rapidly we take a piece of smooth clock mainspring about 3/4" and 10" long and double it together, softening the bending point with the lamp until the piece of mainspring assumes the form shown at Fig. 42, where c represents the piece of spring and $H H$ the bench-vise jaws. The piece of soft steel is placed between the limbs of c c' of the old mainspring up to the line a, Fig. 41, and clamped in the vise jaws. The superfluous steel is cut away with a sharp and rather thin cold chisel.

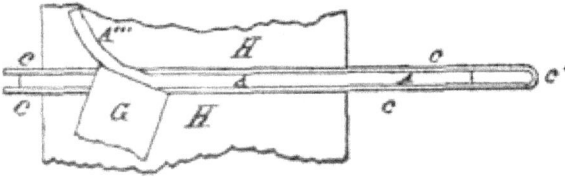

Fig. 42

The chisel is presented as shown at G, Fig. 43 (which is an end view of the vise jaws $H H$ and regulator bar), and held to cut obliquely and with a sort of shearing action, as illustrated in Fig. 42, where A'' represents the soft steel and G the cold chisel. We might add that Fig. 42 is a view of Fig. 43 seen in the direction of the arrow f. It is well to cut in from the edge b on the line d, Fig. 41, with a saw, in order to readily break out the surplus steel and not bend the regulator bar. By setting the pieces of steel obliquely in the vise, or so the line e comes even with the vise jaws, we can cut to more nearly conform to the circular loop A'' of the regulator A.

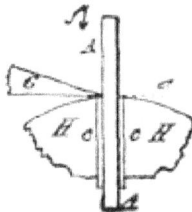

Fig. 43

The smooth steel surface of the bent mainspring c prevents the vise jaws from marking the soft steel of the regulator bar. A person

51

who has not tried this method of cutting out soft steel would not believe with what facility pieces can be shaped. Any workman who has a universal face plate to his lathe can turn out the center of the regulator bar to receive the disk C, and also turn out the center of the regulator spring B. What we have said about the regulator bar applies also to the regulator spring B. This spring is attached to the cock D by means of two small screws at n.

The micrometer screw F is tapped through B'' as in the ordinary Howard regulator, and the screw should be about No. 6 of a Swiss screw-plate. The wire from which such screw is made should be 1/10" in diameter. The steel cap C is fitted like the finer forms of Swiss watches. The hairspring stud E is of steel, shaped as shown, and comes outlined with the other parts.

TO TEMPER AND POLISH STEEL.

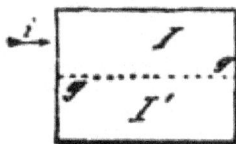

Fig. 44

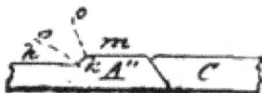

Fig. 46

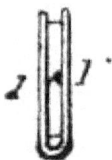

Fig. 45

The regulator bar should be hardened by being placed in a folded piece of sheet iron and heated red hot, and thrown into cold water. The regulator bar A A' is about 3" long; and for holding it for hardening, cut a piece of thin sheet iron 2-1/2" by 3-1/4" and fold it through the middle lengthwise, as indicated by the dotted line g, Fig. 44. The sheet iron when folded will appear as shown at Fig. 45. A piece of flat sheet metal of the same thickness as the regulator bar should be placed between the iron leaves I I, and the leaves beaten down with a hammer, that the iron may serve as a support for the regulator during heating and hardening. A paste made of castile soap and water applied to the regulator bar in the iron envelope will protect it from oxidizing much during the heating. The portions of the regulator bar marked h are intended to be rounded, while the parts marked m are intended to be dead flat. The rounding is carefully done, first with a file and finished with emery paper. The outer edge of the loop A'' is a little rounded, also the inner edge next the cap C. This will be understood by inspecting Fig. 46, where we show a magnified vertical section of the regulator on line l, Fig. 40. The curvature should embrace that portion of A'' between the radial lines o o', and should, on the model, not measure more than 1/40". It will be seen that the curved surface of the regulator is sunk so it meets only the vertical edge of the loop A''. For the average workman, polishing the flat parts m is the most difficult to do, and for this reason we will give entire details. It is to be expected that the regulator bar will spring a little in hardening, but if only a little we need pay no attention to it.

HOW FLAT STEEL POLISHING IS DONE.

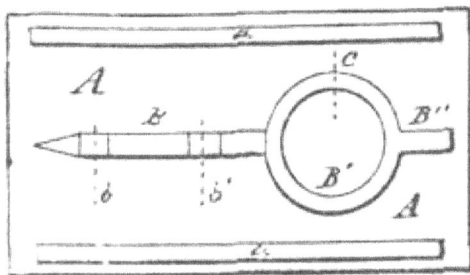

Fig. 47

53

Fig. 48

    Polishing a regulator bar for a large model, such as we are building, is only a heavy job of flat steel work, a little larger but no more difficult than to polish a regulator for a sixteen-size watch. We would ask permission here to say that really nice flat steel work is something which only a comparatively few workmen can do, and, still, the process is quite simple and the accessories few and inexpensive. First, ground-glass slab 6" by 6" by 1/4"; second, flat zinc piece 3-1/4" by 3-1/4" by 1/4"; third, a piece of thick sheet brass 3" by 2" by 1/8"; and a bottle of Vienna lime. The glass slab is only a piece of plate glass cut to the size given above. The zinc slab is pure zinc planed dead flat, and the glass ground to a dead surface with another piece of plate glass and some medium fine emery and water, the whole surface being gone over with emery and water until completely depolished. The regulator bar, after careful filing and dressing up on the edges with an oilstone slip or a narrow emery buff, is finished as previously described. We would add to the details already given a few words on polishing the edges.

Fig. 49

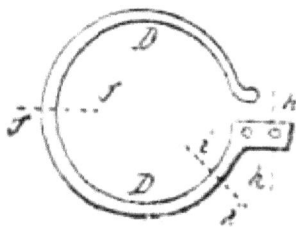

Fig. 50

It is not necessary that the edges of steelwork, like the regulator bar *B*, Fig. 47, should be polished to a flat surface; indeed, they look better to be nicely rounded. Perhaps we can convey the idea better by referring to certain parts: say, spring to the regulator, shown at *D*, Fig. 40, and also the hairspring stud *E*. The edges of these parts look best beveled in a rounded manner.

It is a little difficult to convey in words what is meant by "rounded" manner. To aid in understanding our meaning, we refer to Figs. 48 and 49, which are transverse sections of *D*, Fig. 50, on the line *f*. The edges of *D*, in Fig. 48, are simply rounded. There are no rules for such rounding—only good judgment and an eye for what looks well. The edges of *D* as shown in Fig. 49 are more on the beveled order. In smoothing and polishing such edges, an ordinary jeweler's steel burnish can be used.

SMOOTHING AND POLISHING.

Fig. 51

The idea in smoothing and polishing such edges is to get a fair gloss without much attention to perfect form, inasmuch as it is the flat surface *d* on top which produces the impression of fine finish. If this is flat and brilliant, the rounded edges, like *g c* can really have quite an inferior polish and still look well. For producing the flat polish on the upper surface of the regulator bar B and spring D, the flat surface d, Figs. 48, 49, 51 and 52, we must attach the regulator bar to a plate of heavy brass, as shown at Fig. 47, where A represents the brass plate, and B the regulator bar, arranged for grinding and polishing flat.

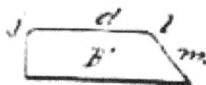

Fig. 52

For attaching the regulator bar B to the brass plate A, a good plan is to cement it fast with lathe wax; but a better plan is to make the plate A of heavy sheet iron, something about 1/8" thick, and secure the two together with three or four little catches of soft solder. It is to be understood the edges of the regulator bar or the regulator spring are polished, and all that remains to be done is to grind and polish the flat face.

Two pieces a a of the same thickness as the regulator bar are placed as shown and attached to A to prevent rocking. After B is securely attached to A, the regulator should be coated with shellac dissolved in alcohol and well dried. The object of this shellac coating is to keep the angles formed at the meeting of the face and side clean in the process of grinding with oilstone dust and oil. The face of the regulator is now placed on the ground glass after smearing it with oil and oilstone dust. It requires but a very slight coating to do the work.

The grinding is continued until the required surface is dead flat, after which the work is washed with soap and water and the shellac dissolved away with alcohol. The final polish is obtained on the zinc lap with Vienna lime and alcohol. Where lathe cement is used for securing the regulator to the plate A, the alcohol used with the Vienna lime dissolves the cement and smears the steel. Diamantine and oil are the best materials for polishing when the regulator bar is cemented to the plate A.

KNOWLEDGE THAT IS MOST ESSENTIAL.

The knowledge most important for a practical working watchmaker to possess is how to get the watches he has to repair in a shape to give satisfaction to his customers. No one will dispute the truth of the above italicised statement. It is only when we seek to have limits set, and define what such knowledge should consist of, that disagreement occurs.

One workman who has read Grossmann or Saunier, or both, would insist on all watches being made to a certain standard, and, according to their ideas, all such lever watches as we are now dealing with should have club-tooth escapements with equidistant lockings, ten degrees lever and pallet action, with one and one-half degrees lock and one and one-half degrees drop. Another workman would insist on circular pallets, his judgment being based chiefly on what he had read as stated by some author. Now the facts of the situation are that lever escapements vary as made by different manufacturers, one concern using circular pallets and another using pallets with equidistant lockings.

WHAT A WORKMAN SHOULD KNOW TO REPAIR A WATCH.

One escapement maker will divide the impulse equally between the tooth and pallet; another will give an excess to the tooth. Now while these matters demand our attention in the highest degree in a theoretical sense, still, for such "know hows" as count in a workshop, they are of but trivial importance in practice.

We propose to deal in detail with the theoretical consideration of "thick" and "thin" pallets, and dwell exhaustively on circular pallets and those with equidistant locking faces; but before we do so we wish to impress on our readers the importance of being able to free themselves of the idea that all lever escapements should conform to the rigid rules of any dictum.

EDUCATE THE EYE TO JUDGE OF ANGULAR AS WELL AS LINEAR EXTENT.

For illustration: It would be easy to design a lever escapement that would have locking faces which were based on the idea of employing neither system, but a compromise between the two, and still give a good, sound action. All workmen should learn to estimate accurately the extent of angular motion, so as to be able to judge correctly of escapement actions. It is not only necessary to know that a club-tooth escapement should have one and one-half degrees drop, but the eye should be educated, so to speak, as to be able to judge of angular as well as linear extent.

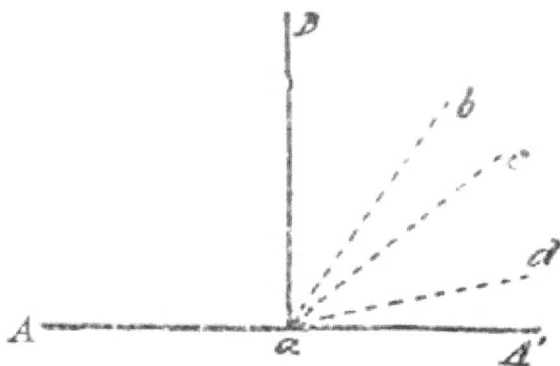

Fig. 53

Most mechanics will estimate the size of any object measured in inches or parts of inches very closely; but as regards angular extent, except in a few instances, we will find mechanics but indifferent judges. To illustrate, let us refer to Fig. 53. Here we have the base line *A A'* and the perpendicular line *a B*. Now almost any person would be able to see if the angle *A a B* was equal to *B a A'*; but not five in one hundred practical mechanics would be able to estimate with even tolerable accuracy the measure the angles made to the base by the lines *b c d*; and still watchmakers are required in the daily practice of their craft to work to angular motions and movements almost as important as to results as diameters.

58

What is the use of our knowing that in theory an escape-wheel tooth should have one and one-half degrees drop, when in reality it has three degrees? It is only by educating the eye from carefully-made drawings; or, what is better, constructing a model on a large scale, that we can learn to judge of proper proportion and relation of parts, especially as we have no convenient tool for measuring the angular motion of the fork or escape wheel. Nor is it important that we should have, if the workman is thoroughly "booked up" in the principles involved.

As we explained early in this treatise, there is no imperative necessity compelling us to have the pallets and fork move through ten degrees any more than nine and one-half degrees, except that experience has proven that ten degrees is about the right thing for good results. In this day, when such a large percentage of lever escapements have exposed pallets, we can very readily manipulate the pallets to match the fork and roller action. For that matter, in many instances, with a faulty lever escapement, the best way to go about putting it to rights is to first set the fork and roller so they act correctly, and then bring the pallets to conform to the angular motion of the fork so adjusted.

FORK AND ROLLER ACTION.

Although we could say a good deal more about pallets and pallet action, still we think it advisable to drop for the present this particular part of the lever escapement and take up fork and roller action, because, as we have stated, frequently the fork and roller are principally at fault. In considering the action and relation of the parts of the fork and roller, we will first define what is considered necessary to constitute a good, sound construction where the fork vibrates through ten degrees of angular motion and is supposed to be engaged with the roller by means of the jewel pin for thirty degrees of angular motion of the balance.

There is no special reason why thirty degrees of roller action should be employed, except that experience in practical construction has come to admit this as about the right arc for watches of ordinary good, sound construction. Manufacturers have made departures from this standard, but in almost every instance have finally come back to pretty near these proportions. In deciding on the length of fork and size of roller, we first decide on the distance apart at which to place the center of the balance and the center of the pallet staff. These two points established, we have the length of the fork and diameter of the roller defined at once.

HOW TO FIND THE ROLLER DIAMETER FROM THE LENGTH OF THE FORK.

To illustrate, let us imagine the small circles A B, Fig. 54, to represent the center of a pallet staff and balance staff in the order named. We divide this space into four equal parts, as shown, and the third space will represent the point at which the pitch circles of the fork and roller will intersect, as shown by the arc a and circle b. Now if the length of the radii of these circles stand to each other as three to one, and the fork vibrates through an arc of ten degrees, the jewel pin engaging such fork must remain in contact with said fork for thirty degrees of angular motion of the balance.

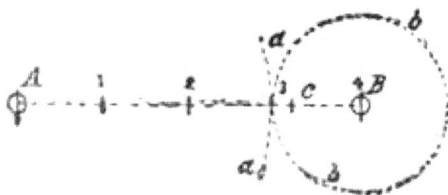

Fig. 54

Or, in other words, the ratio of angular motion of two *mobiles* acting on each must be in the same ratio as the length of their radii at the point of contact. If we desire to give the jewel pin, or, in ordinary horological phraseology, have a greater arc of roller action, we would extend the length of fork (say) to the point c, which would be one-fifth of the space between A and B, and the ratio of fork to roller action would be four to one, and ten degrees of fork action would give forty degrees of angular motion to the roller—and such escapements have been constructed.

60

WHY THIRTY DEGREES OF ROLLER ACTION IS ABOUT RIGHT.

Now we have two sound reasons why we should not extend the arc of vibration of the balance: (*a*) If there is an advantage to be derived from a detached escapement, it would surely be policy to have the arc of contact, that is, for the jewel pin to engage the fork, as short an arc as is compatible with a sound action. (*b*) It will be evident to any thinking mechanic that the acting force of a fork which would carry the jewel pin against the force exerted by the balance spring through an arc of fifteen degrees, or half of an arc of thirty degrees, would fail to do so through an arc of twenty degrees, which is the condition imposed when we adopt forty degrees of roller action.

For the present we will accept thirty degrees of roller action as the standard. Before we proceed to delineate our fork and roller we will devote a brief consideration to the size and shape of a jewel pin to perform well. In this matter there has been a broad field gone over, both theoretically and in practical construction. Wide jewel pins, round jewel pins, oval jewel pins have been employed, but practical construction has now pretty well settled on a round jewel pin with about two-fifths cut away. And as regards size, if we adopt the linear extent of four degrees of fork or twelve degrees of roller action, we will find it about right.

HOW TO SET A FORK AND ROLLER ACTION RIGHT.

As previously stated, frequently the true place to begin to set a lever escapement right is with the roller and fork. But to do this properly we should know when such fork and roller action is right and safe in all respects. We will see on analysis of the actions involved that there are three important actions in the fork and roller functions: (*a*) The fork imparting perfect impulse through the jewel pin to the balance. (*b*) Proper unlocking action. (*c*) Safety action. The last function is in most instances sadly neglected and, we regret to add, by a large majority of even practical workmen it is very imperfectly understood. In most American watches we have ample opportunity afforded to inspect the pallet action, but the fork and roller action is placed so that rigid inspection is next to impossible.

The Vacheron concern of Swiss manufacturers were acute enough to see the importance of such inspection, and proceeded to cut a circular opening in the lower plate, which permitted, on the removal of the dial, a careful scrutiny of the action of the roller and fork. While writing on this topic we would suggest the importance not only of knowing how to draw a correct fork and roller action, but letting the workman who desires to be *au fait* in escapements delineate and study the action of a faulty fork and roller action—say one in which the fork, although of the proper form, is too short, or what at first glance would appear to amount to the same thing, a roller too small.

Drawings help wonderfully in reasoning out not only correct actions, but also faulty ones, and our readers are earnestly advised to make such faulty drawings in several stages of action. By this course they will educate the eye to discriminate not only as to correct actions, but also to detect those which are imperfect, and we believe most watchmakers will admit that in many instances it takes much longer to locate a fault than to remedy it after it has been found.

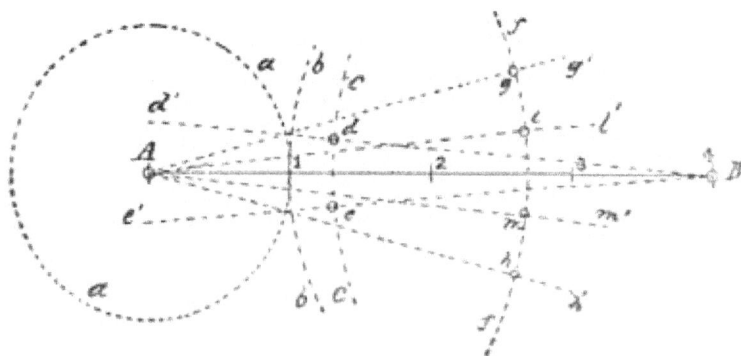

Fig. 55

Let us now proceed to delineate a fork and roller. It is not imperative that we should draw the parts to any scale, but it is a rule among English makers to let the distance between the center of the pallet staff and the center of the balance staff equal in length the chord of ninety-six degrees of the pitch circle of the escape wheel, which, in case we employ a pitch circle of 5" radius, would make the distance between A and B, Fig. 55, approximately 7-1/2", which is a very fair scale for study drawings.

HOW TO DELINEATE A FORK AND ROLLER.

To arrive at the proper proportions of the several parts, we divide the space *A B* into four equal parts, as previously directed, and draw the circle *a* and short arc *b*. With our dividers set at 5", from *B* as a center we sweep the short arc *c*. From our arc of sixty degrees, with a 5" radius, we take five degrees, and from the intersection of the right line *A B* with the arc *c* we lay off on each side five degrees and establish the points *d e*; and from *B* as a center, through these points draw the lines *B d'* and *B e'*. Now the arc embraced between these lines represents the angular extent of our fork action.

From *A* as a center and with our dividers set at 5", we sweep the arc *f*. From the scale of degrees we just used we lay off fifteen degrees on each side of the line *A B* on the arc *f*, and establish the points *g h*. From *A* as a center, through the points just established we draw the radial lines *A g'* and *A h'*. The angular extent between these lines defines the limit of our roller action.

Now if we lay off on the arc *f* six degrees each side of its intersection with the line *A B*, we define the extent of the jewel pin; that is, on the arc *f* we establish the points *l m* at six degrees from the line *A B*, and through the points *l m* draw, from *A* as a center, the radial lines *A l'* and *A m'*. The extent of the space between the lines *A l'* and *A m'* on the circle *a* defines the size of our jewel pin.

TO DETERMINE THE SIZE OF A JEWEL PIN.

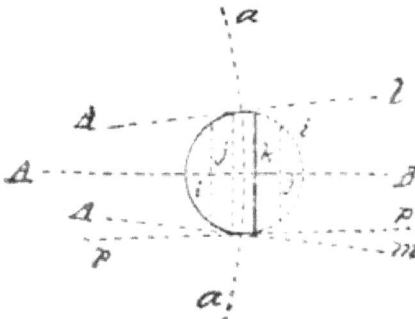

Fig. 56

To make the situation better understood, we make an enlarged drawing of the lines defining the jewel pin at Fig. 56. At the intersection of the line A B with the arc a we locate the point k, and from it as a center we sweep the circle i so it passes through the intersection of the lines A l' and A m' with the arc a. We divide the radius of the circle i on the line A B into five equal parts, as shown by the vertical lines j. Of these five spaces we assume three as the extent of the jewel pin, cutting away that portion to the right of the heavy vertical line at k.

We will now proceed to delineate a fork and roller as the parts are related on first contact of jewel pin with fork and initial with the commencing of the act of unlocking a pallet. The position and relations are also the same as at the close of the act of impulse. We commence the drawing at Fig. 57, as before, by drawing the line A B and the arcs a and b to represent the pitch circles. We also sweep the arc f to enable us to delineate the line A g'. Next in order we draw our jewel pin as shown at D. In drawing the jewel pin we proceed as at Fig. 56, except we let the line A g', Fig. 57, assume the same relations to the jewel pin as A B in Fig. 56; that is, we delineate the jewel pin as if extending on the arc a six degrees on each side of the line A g', Fig. 57.

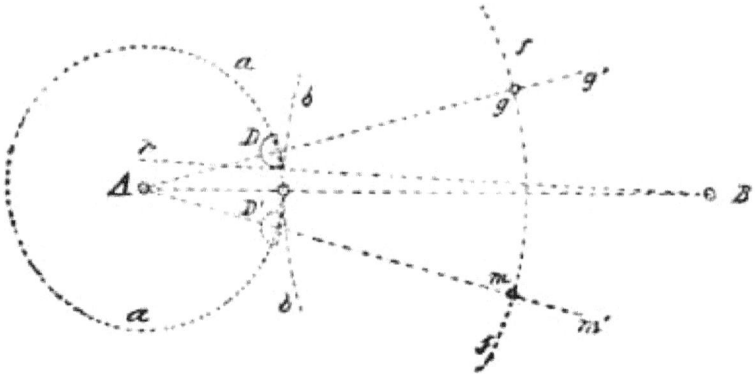

Fig. 57

THE THEORY OF THE FORK ACTION.

To aid us in reasoning, we establish the point m, as in Fig. 55, at m, Fig. 57, and proceed to delineate another and imaginary jewel pin at D' (as we show in dotted outline). A brief reasoning will show that in allowing thirty degrees of contact of the fork with the jewel pin, the center of the jewel pin will pass through an arc of thirty degrees, as shown on the arcs a and f. Now here is an excellent opportunity to impress on our minds the true value of angular motion, inasmuch as thirty degrees on the arc f is of more than twice the linear extent as on the arc a.

Before we commence to draw the horn of the fork engaging the jewel pin D, shown at full line in Fig. 57, we will come to perfectly understand what mechanical relations are required. As previously stated, we assume the jewel pin, as shown at D, Fig. 57, is in the act of encountering the inner face of the horn of the fork for the end or purpose of unlocking the engaged pallet. Now if the inner face of the horn of the fork was on a radial line, such radial line would be $p\ B$, Fig. 57. We repeat this line at p, Fig. 56, where the parts are drawn on a larger scale.

To delineate a fork at the instant the last effort of impulse has been imparted to the jewel pin, and said jewel pin is in the act of separating from the inner face of the prong of the fork—we would also call attention to the fact that relations of parts are precisely the same as if the jewel pin had just returned from an excursion of vibration and was in the act of encountering the inner face of the prong of the fork in the act of unlocking the escapement.

We mentioned this matter previously, but venture on the repetition to make everything clear and easily understood. We commence by drawing the line $A\ B$ and dividing it in four equal parts, as on previous occasions, and from A and B as centers draw the pitch circles c d. By methods previously described, we draw the lines $A\ a$ and $A\ a'$, also $B\ b$ and $B\ b'$ to represent the angular motion of the two mobiles, viz., fork and roller action. As already shown, the roller occupies twelve degrees of angular extent. To get at this conveniently, we lay off on the arc by which we located the lines $A\ a$ and $A\ a'$ six degrees above the line $A\ a$ and draw the line $A\ h$.

Now the angular extent on the arc c between the lines $A\ a$ and $A\ h$ represents the radius of the circle defining the jewel pin. From the intersection of the line $A\ a$ with the arc c as a center, and with the radius just named, we sweep the small circle D, Fig. 58, which represents our jewel pin; we afterward cut away two-fifths and draw the full line D, as shown. We show at Fig. 59 a portion of Fig. 58, enlarged four times, to show certain portions of our delineations more distinctly. If we give the subject a moment's consideration we will see that the length of the prong E of the lever fork is limited to such a length as will allow the jewel pin D to pass it.

HOW TO DELINEATE THE PRONGS OF A LEVER FORK.

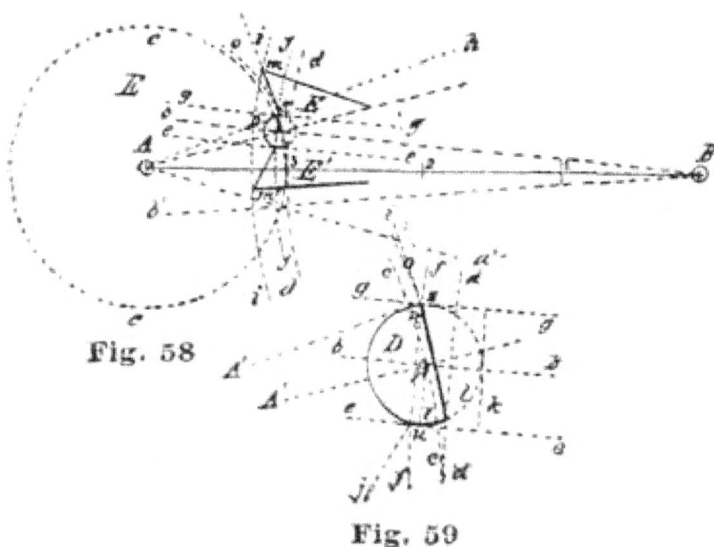

Fig. 58

Fig. 59

To delineate this length, from B as a center we sweep the short arc f so it passes through the outer angle n, Fig. 59, of the jewel pin. This arc, carried across the jewel pin D, limits the length of the opposite prong of the fork. The outer face of the prong of the fork can be drawn as a line tangent to a circle drawn from A as a center through the angle n of the jewel pin. Such a circle or arc is shown at o, Figs. 58 and 59. There has been a good deal said as to whether the outer edge of the prong of a fork should be straight or curved.

To the writer's mind, a straight-faced prong, like from s to m, is what is required for a fork with a single roller, while a fork with a

66

curved prong will be best adapted for a double roller. This subject will be taken up again when we consider double-roller action. The extent or length of the outer face of the prong is also an open subject, but as there is but one factor of the problem of lever escapement construction depending on it, when we name this and see this requirement satisfied we have made an end of this question. The function performed by the outer face of the prong of a fork is to prevent the engaged pallet from unlocking while the guard pin is opposite to the passing hollow.

The inner angle *s* of the horn of the fork must be so shaped and located that the jewel pin will just clear it as it passes out of the fork, or when it passes into the fork in the act of unlocking the escapement. In escapements with solid bankings a trifle is allowed, that is, the fork is made enough shorter than the absolute theoretical length to allow for safety in this respect.

THE PROPER LENGTH OF A LEVER.

We will now see how long a lever must be to perform its functions perfectly. Now let us determine at what point on the inner face of the prong *E'* the jewel pin parts from the fork, or engages on its return. To do this we draw a line from the center *r* (Fig. 59) of the jewel pin, so as to meet the line *e* at right angles, and the point *t* so established on the line *e* is where contact will take place between the jewel pin and fork.

It will be seen this point (*t*) of contact is some distance back of the angle *u* which terminates the inner face of the prong *E'*; consequently, it will be seen the prongs *E E'* of the fork can with safety be shortened enough to afford a safe ingress or egress to the jewel pin to the slot in the fork. As regards the length of the outer face of the prong of the fork, a good rule is to make it one and a half times the diameter of the jewel pin. The depth of the slot need be no more than to free the jewel in its passage across the ten degrees of fork action. A convenient rule as to the depth of the slot in a fork is to draw the line *k*, which, it will be seen, coincides with the circle which defines the jewel pin.

67

HOW TO DELINEATE THE SAFETY ACTION.

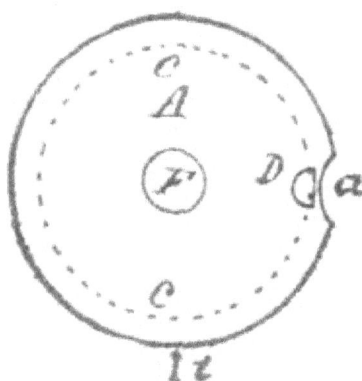

Fig. 60

We will next consider a safety action of the single roller type. The active or necessary parts of such safety action consist of a roller or disk of metal, usually steel, shaped as shown in plan at *A*, Fig. 60. In the edge of this disk is cut in front of the jewel pin a circular recess shown at *a* called the passing hollow. The remaining part of the safety action is the guard pin shown at *N* Figs. 61 and 62, which is placed in the lever. Now it is to be understood that the sole function performed by the guard pin is to strike the edge of the roller *A* at any time when the fork starts to unlock the engaged pallet, except when the jewel pin is in the slot of the fork. To avoid extreme care in fitting up the pass-ing hollow, the horns of the fork are arranged to strike the jewel pin and prevent unlocking in case the passing hollow is made too wide. To delineate the safety action we first draw the fork and jewel pin as pre-viously directed and as shown at Fig. 63. The position of the guard pin should be as close to the bottom of the slot of the fork as possible and be safe. As to the size of the guard pin, it is usual to make it about one-third or half the diameter of the jewel pin. The size and position of the guard pin decided on and the small circle *N* drawn, to define the size and position of the roller we set our dividers so that a circle drawn from the center *A* will just touch the edge of the small circle *N*, and thus define the outer boundary of our roller, or roller table, as it is fre-quently called.

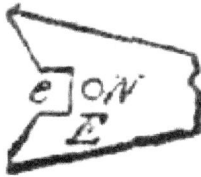

Fig. 61

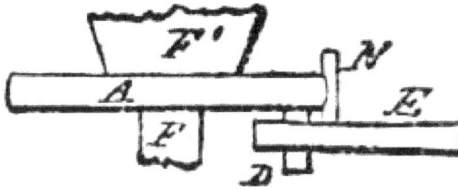

Fig. 62

For deciding the angular extent of the passing hollow we have no fixed rule, but if we make it to occupy about half more angular extent on the circle y than will coincide with the angular extent of the jewel pin, it will be perfectly safe and effectual. We previously stated that the jewel pin should occupy about twelve degrees of angular extent on the circle c, and if we make the passing hollow occupy eighteen degrees (which is one and a half the angular extent of the jewel pin) it will do nicely. But if we should extend the width of the passing hollow to twenty-four degrees it would do no harm, as the jewel pin would be well inside the horn of the fork before the guard pin could enter the passing hollow.

We show in Fig. 61 the fork as separated from the roller, but in Fig. 62, which is a side view, we show the fork and jewel pin as engaged. When drawing a fork and roller action it is safe to show the guard pin as if in actual contact with the roller. Then in actual construction, if the parts are made to measure and agree with the drawing in the gray, that is, before polishing, the process of polishing will reduce the convex edge of the roller enough to free it.

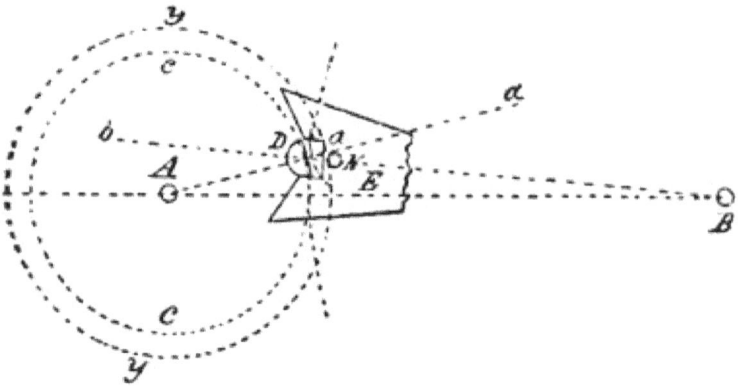

Fig. 63

It is evident if thought is given to the matter, that if the guard pin is entirely free and does not touch the roller in any position, a condition and relation of parts exist which is all we can desire. We are aware that it is usual to give a considerable latitude in this respect even by makers, and allow a good bit of side shake to the lever, but our judgment would condemn the practice, especially in high-grade watches.

RESTRICT THE FRICTIONAL SURFACES.

Grossmann, in his essay on the detached lever escapement, adopts one and a half degrees lock. Now, we think that one degree is ample; and we are sure that every workman experienced in the construction of the finer watches will agree with us in the assertion that we should in all instances seek to reduce the extent of all frictional surfaces, no matter how well jeweled. Acting under such advice, if we can reduce the surface friction on the lock from one and a half degrees to one degree or, better, to three-fourths of a degree, it is surely wise policy to do so. And as regards the extent of angular motion of the lever, if we reduce this to six degrees, exclusive of the lock, we would undoubtedly obtain better results in timing.

We shall next consider the effects of opening the bankings too wide, and follow with various conditions which are sure to come in the experience of the practical watch repairer. It is to be supposed in this problem that the fork and roller action is all right. The reader may say to this, why not close the banking? In reply we would offer the supposition that some workman had bent the guard pin forward or set a pallet stone too far out.

We have now instructed our readers how to draw and construct a lever escapement complete, of the correct proportions, and will next take up defective construction and consider faults existing to a lesser or greater degree in almost every watch. Faults may also be those arising from repairs by some workman not fully posted in the correct form and relation of the several parts which go to make up a lever escapement. It makes no difference to the artisan called upon to put a watch in perfect order as to whom he is to attribute the imperfection, maker or former repairer; all the workman having the job in hand has to do is to know positively that such a fault actually exists, and that it devolves upon him to correct it properly.

BE FEARLESS IN REPAIRS, IF SURE YOU ARE RIGHT.

Hence the importance of the workman being perfectly posted on such matters and, knowing that he is right, can go ahead and make the watch as it should be. The writer had an experience of this kind years ago in Chicago. A Jules Jurgensen watch had been in the hands of several good workmen in that city, but it would stop. It was then brought to him with a statement of facts given above. He knew there must be a fault somewhere and searched for it, and found it in the exit pallet—a certain tooth of the escape wheel under the right conditions would sometimes not escape. It might go through a great many thousand times and yet it might, and did sometimes, hold enough to stop the watch.

Now probably most of my fellow-workmen in this instance would have been afraid to alter a "Jurgensen," or even hint to the owner that such a thing could exist as a fault in construction in a watch of this justly-celebrated maker. The writer removed the stone, ground a little from the base of the offending pallet stone, replaced it, and all trouble ended—no stops from that on.

STUDY OF AN ESCAPEMENT ERROR.

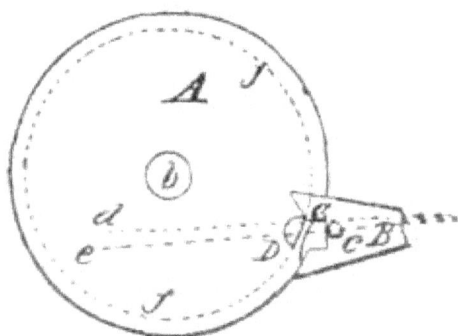

Fig. 64

Now let us suppose a case, and imagine a full-plate American movement in which the ingress or entrance pallet extends out too far, and in order to have it escape, the banking on that side is opened too wide. We show at Fig. 64 a drawing of the parts in their proper relations under the conditions named. It will be seen by careful inspection that the jewel pin D will not enter the fork, which is absolutely necessary. This condition very frequently exists in watches where a new pallet stone has been put in by an inexperienced workman. Now this is one of the instances in which workmen complain of hearing a "scraping" sound when the watch is placed to the ear. The remedy, of course, lies in warming up the pallet arms and pushing the stone in a trifle, "But how much?" say some of our readers. There is no definite rule, but we will tell such querists how they can test the matter.

Remove the hairspring, and after putting the train in place and securing the plates together, give the winding arbor a turn or two to put power on the train; close the bankings well in so the watch cannot escape on either pallet. Put the balance in place and screw down the cock. Carefully turn back the banking on one side so the jewel pin will just pass out of the slot in the fork. Repeat this process with the opposite banking; the jewel pin will now pass out on each side. Be sure the guard pin does not interfere with the fork action in any way. The fork is now in position to conform to the conditions required.

HOW TO ADJUST THE PALLETS TO MATCH THE FORK.

If the escapement is all right, the teeth will have one and a half degrees lock and escape correctly; but in the instance we are considering, the stone will not permit the teeth to pass, and must be pushed in until they will. It is not a very difficult matter after we have placed the parts together so we can see exactly how much the pallet protrudes beyond what is necessary, to judge how far to push it back when we have it out and heated. There is still an "if" in the problem we are considering, which lies in the fact that the fork we are experimenting with may be too short for the jewel pin to engage it for ten degrees of angular motion.

This condition a man of large experience will be able to judge of very closely, but the better plan for the workman is to make for himself a test gage for the angular movement of the fork. Of course it will be understood that with a fork which engages the roller for eight degrees of fork action, such fork will not give good results with pallets ground for ten degrees of pallet action; still, in many instances, a compromise can be effected which will give results that will satisfy the owner of a watch of moderate cost, and from a financial point of view it stands the repairer in hand to do no more work than is absolutely necessary to keep him well pleased.

We have just made mention of a device for testing the angular motion of the lever. Before we take up this matter, however, we will devote a little time and attention to the subject of jewel pins and how to set them. We have heretofore only considered jewel pins of one form, that is, a round jewel pin with two-fifths cut away. We assumed this form from the fact that experience has demonstrated that it is the most practicable and efficient form so far devised or applied. Subsequently we shall take up the subject of jewel pins of different shapes.

HOW TO SET A JEWEL PIN AS IT SHOULD BE.

Many workmen have a mortal terror of setting a jewel pin and seem to fancy that they must have a specially-devised instrument for accomplishing this end. Most American watches have the hole for the jewel pin "a world too wide" for it, and we have heard repeated com-

plaints from this cause. Probably the original object of this accommo-
dating sort of hole was to favor or obviate faults of pallet action. Let us
suppose, for illustration, that we have a roller with the usual style of
hole for a jewel pin which will take almost anything from the size of a
No. 12 sewing needle up to a round French clock pallet.

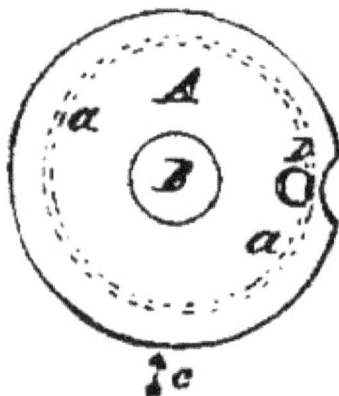

Fig. 65

We are restricted as regards the proper size of jewel pin by the
width of the slot in the fork. Selecting a jewel which just fits the fork,
we can set it as regards its relation to the staff so it will cause the pitch
circle of the jewel pin to coincide with either of dotted circles a or a',
Fig. 65. This will perhaps be better understood by referring to Fig. 66,
which is a view of Fig. 65 seen in the direction of the arrow c. Here we
see the roller jewel at D, and if we bring it forward as far as the hole in
the roller will permit, it will occupy the position indicated at the dotted
lines; and if we set it in (toward the staff) as far as the hole will allow,
it will occupy the position indicated by the full outline.

Fig. 66

Now such other condition might very easily exist, that bringing the jewel pin forward to the position indicated by the dotted lines at *D*, Fig. 66, would remedy the defect described and illustrated at Fig. 64 without any other change being necessary. We do not assert, understand, that a hole too large for the jewel pin is either necessary or desirable—what we wish to convey to the reader is the necessary knowledge so that he can profit by such a state if necessary. A hole which just fits the jewel pin so the merest film of cement will hold it in place is the way it should be; but we think it will be some time before such rollers are made, inasmuch as economy appears to be a chief consideration.

ABOUT JEWEL-PIN SETTERS.

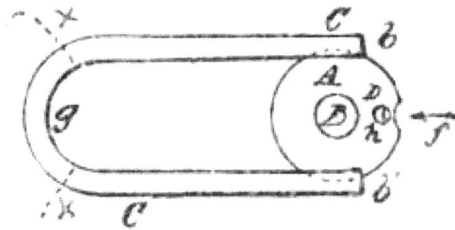

Fig. 67

Fig. 68

To make a jewel-pin setter which will set a jewel pin straight is easy enough, but to devise any such instrument which will set a jewel so as to perfectly accord with the fork action is probably not practicable. What the workman needs is to know from examination when the jewel pin is in the proper position to perform its functions correctly, and he can only arrive at this knowledge by careful study and thought on the matter. If we make up our minds on examining a watch that a jewel pin is "set too wide," that is, so it carries the fork over too far and increases the lock to an undue degree, take out the balance, remove the hairspring, warm the roller with a small alcohol lamp, and then with the tweezers move the jewel pin in toward the staff.

Fig. 69

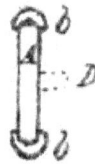

Fig. 70

No attempt should be made to move a jewel pin unless the cement which holds the jewel is soft, so that when the parts cool off the jewel is as rigid as ever. A very little practice will enable any workman who has the necessary delicacy of touch requisite to ever become a good watchmaker, to manipulate a jewel pin to his entire satisfaction with no other setter than a pair of tweezers and his eye, with a proper knowledge of what he wants to accomplish. To properly heat a roller for truing up the jewel pin, leave it on the staff, and after removing the hairspring hold the balance by the rim in a pair of tweezers, "flashing it" back and forth through the flame of a rather small alcohol lamp until the rim of the balance is so hot it can just be held between the thumb and finger, and while at this temperature the jewel pin can be pressed forward or backward, as illustrated in Fig. 66, and then a touch or two will set the pin straight or parallel with the staff. Figs. 68 and 69 are self-explanatory. For cementing in a jewel pin a very convenient tool is shown at Figs. 67 and 70. It is made of a piece of copper wire about 1/16" in diameter, bent to the form shown at Fig. 67. The ends *b b* of the copper wire are flattened a little and recessed on their inner faces, as shown in Fig. 70, to grasp the edges of the roller *A*. The heat of an alcohol lamp is applied to the loop of the wire at *g* until the small bit of shellac placed in the hole *h* melts. The necessary small pieces of shellac are made by warming a bit of the gum to near the melting point and then drawing the softened gum into a filament the size of horse hair. A bit of this broken off and placed in the hole *h* supplies the cement necessary to fasten the jewel pin. Figs. 68 and 69 will, no doubt, assist in a clear understanding of the matter.

HOW TO MAKE AN ANGLE-MEASURING DEVICE.

We will now resume the consideration of the device for measuring the extent of the angular motion of the fork and pallets. Now, before we take this matter up in detail we wish to say, or rather repeat what we have said before, which is to the effect that ten degrees of fork and lever action is not imperative, as we can get just as sound an action and precisely as good results with nine and a half or even nine degrees as with ten, if other acting parts are in unison with such an arc of angular motion. The chief use of such an angle-measuring device is to aid in comparing the relative action of the several parts with a known standard.

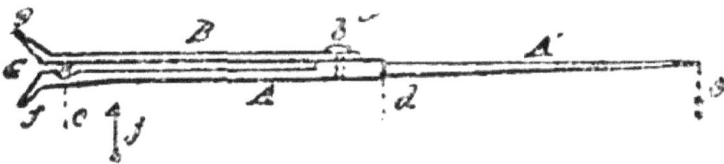

Fig. 71

For use with full-plate movements about the best plan is a spring clip or clasp to embrace the pallet staff below the pallets. We show at Fig. 71 such a device. To make it, take a rather large size of sewing needle—the kind known as a milliner's needle is about the best. The diameter of the needle should be about No. 2, so that at *b* we can drill and put in a small screw. It is important that the whole affair should be very light. The length of the needle should be about 1-5/8", in order that from the notch a to the end of the needle A' should be 1-1/2". The needle should be annealed and flattened a little, to give a pretty good grasp to the notch a on the pallet staff.

Good judgment is important in making this clamp, as it is nearly impossible to give exact measurements. About 1/40" in width when seen in the direction of the arrow j will be found to be about the right width. The spring B can be made of a bit of mainspring, annealed and filed down to agree in width with the part A. In connection with the device shown at Fig. 71 we need a movement-holder to hold the movement as nearly a constant height as possible above the bench. The idea is, when the clamp A B is slipped on the pallet staff the index hand A' will extend outward, as shown in Fig. 72, where the circle C is supposed to represent the top plate of a watch, and A' the index hand.

HOW THE ANGULAR MOTION IS MEASURED.

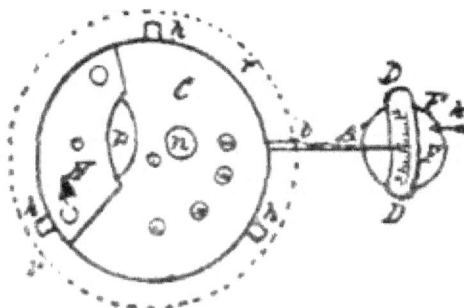

Fig. 72

Fig. 72 is supposed to be seen from above. It is evident that if we remove the balance from the movement shown at C, leaving power on the train, and with an oiling tool or hair broach move the lever back and forth, the index hand A' will show in a magnified manner the angular motion of the lever. Now if we provide an index arc, as shown at D, we can measure the extent of such motion from bank to bank.

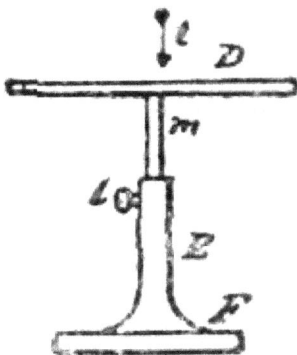

Fig. 73

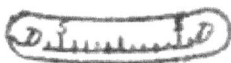

Fig. 74

To get up such an index arc we first make a stand as shown at *E F*, Fig. 73. The arc *D* is made to 1-1/2" radius, to agree with the index hand *A'*, and is divided into twelve degree spaces, six each side of a zero, as shown at Fig. 74, which is an enlarged view of the index *D* in Fig. 72. The index arc is attached to a short bit of wire extending down into the support *E*, and made adjustable as to height by the set-screw *l*. Let us suppose the index arc is adjusted to the index hand *A'*, and we move the fork as suggested; you see the hand would show exactly the arc passed through from bank to bank, and by moving the stand *E F* we can arrange so the zero mark on the scale stands in the center of such arc. This, of course, gives the angular motion from bank to bank. As an experiment, let us close the bankings so they arrest the fork at the instant the tooth drops from each pallet. If this arc is ten degrees, the pallet action is as it should be with the majority of modern watches.

TESTING LOCK AND DROP WITH OUR NEW DEVICE.

Let us try another experiment: We carefully move the fork away from the bank, and if after the index hand has passed through one and a half degrees the fork flies over, we know the lock is right. We repeat the experiment from the opposite bank, and in the same manner determine if the lock is right on the other pallets. You see we have now the means of measuring not only the angular motion of the lever, but the angular extent of the lock. At first glance one would say that if now we bring the roller and fork action to coincide and act in unison with the pallet action, we would be all right; and so we would, but frequently this bringing of the roller and fork to agree is not so easily accomplished.

It is chiefly toward this end the Waltham fork is made adjustable, so it can be moved to or from the roller, and also that we can allow the pallet arms to be moved, as we will try and explain. As we set the bankings the pallets are all right; but to test matters, let us remove the hairspring and put the balance in place. Now, if the jewel pin passes in and out of the fork, it is to be supposed the fork and roller action is all right. To test the fork and roller action we close the banking a little on one side. If the fork and jewel pin are related to each other as they should be, the jewel pin will not pass out of the fork, nor will the engaged tooth drop from that pallet. This condition should obtain on both pallets, that is, if the jewel pin will not pass out of the fork on a given bank the tooth engaged on its pallet should not drop.

We have now come to the most intricate and important problems which relate to the lever escapement. However, we promise our readers that if they will take the pains to follow closely our elucidations, to make these puzzles plain. But we warn them that they are no easy problems to solve, but require good, hard thinking. The readiest way to master this matter is by means of such a model escapement as we have described. With such a model, and the pallets made to clamp with small set-screws, and roller constructed so the jewel pin could be set to or from the staff, this matter can be reduced to object lessons. But study of the due relation of the parts in good drawings will also master the situation.

A FEW EXPERIMENTS WITH OUR ANGLE-MEASURING DEVICE.

In using the little instrument for determining angular motion that we have just described, care must be taken that the spring clamp which embraces the pallet staff does not slip. In order to thoroughly understand the methods of using this angle-measuring device, let us take a further lesson or two.

We considered measuring the amount of lock on each pallet, and advised the removal of the balance, because if we left the balance in we could not readily tell exactly when the tooth passed on to the impulse plane; but if we touch the fork lightly with an oiling tool or a hair broach, moving it (the fork) carefully away from the bank and watching the arc indicated by the hand *A*, Fig. 72, we can determine with great exactness the angular extent of lock. The diagram at Fig. 75 illustrates how this experiment is conducted. We apply the hair broach to the end of the fork *M*, as shown at *L*, and gently move the fork in the direction of the arrow *i*, watching the hand *A* and note the number of degrees, or parts of degrees, indicated by the hand as passed over before the tooth is unlocked and passes on to the impulse plane and the fork flies forward to the opposite bank. Now, the quick movement of the pallet and fork may make the hand mark more or less of an arc on the index than one of ten degrees, as the grasp may slip on the pallet staff; but the arc indicated by the slow movement in unlocking will be correct.

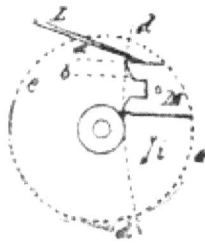

Fig. 75

By taking a piece of sharpened pegwood and placing the point in the slot of the fork, we can test the fork to see if the drop takes place much before the lever rests against the opposite bank. As we have previously stated, the drop from the pallet should not take place until the lever *almost* rests on the banking pin. What the reader should impress on his mind is that the lever should pass through about one and a half degrees arc to unlock, and the remainder (eight and a half degrees) of the ten degrees are to be devoted to impulse. But, understand, if the

81

impulse angle is only seven and a half degrees, and the jewel pin acts in accordance with the rules previously given, do not alter the pallet until you know for certain you will gain by it. An observant workman will, after a little practice, be able to determine this matter.

We will next take up the double roller and fork action, and also consider in many ways the effect of less angles of action than ten degrees. This matter now seems of more importance, from the fact that we are desirous to impress on our readers that *there is no valid reason for adopting ten degrees of fork and roller action with the table roller, except that about this number of degrees of action are required to secure a reliable safety action.* With the double roller, as low as six degrees fork and pallet action can be safely employed. In fork and pallet actions below six degrees of angular motion, side-shake in pivot holes becomes a dangerous factor, as will be explained further on. It is perfectly comprehending the action of the lever escapement and then being able to remedy defects, that constitute the master workman.

HOW TO MEASURE THE ANGULAR MOTION OF AN ESCAPE WHEEL.

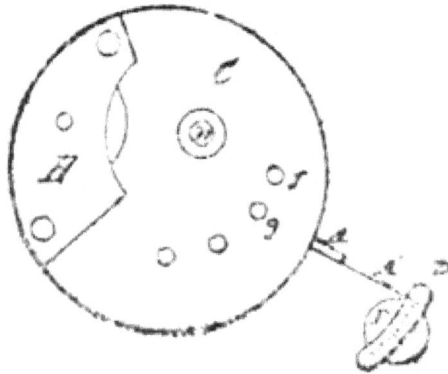

Fig. 76

We can also make use of our angle-testing device for measuring our escape-wheel action, by letting the clasp embrace the arbor of the escape wheel, instead of the pallet staff. We set the index arc as in our former experiments, except we place the movable index *D*, Fig. 76, so that when the engaged tooth rests on the locking face of a pallet, the index hand stands at the extreme end of our arc of twelve degrees. We next, with our pointed pegwood, start to move the fork away from the bank, as before, we look sharp and see the index hand move backward a little, indicating the "draw" on the locking face. As soon as the pallet reaches the impulse face, the hand *A* moves rapidly forward, and if the escapement is of the club-tooth order and closely matched, the hand *A* will pass over ten and a half degrees of angular motion before the drop takes place.

Fig. 77

We will warn our readers in advance, that if they make such a testing device they will be astonished at the inaccuracy which they will find in the escapements of so-called fine watches. The lock, in many instances, instead of being one and a half degrees, will oftener be found to be from two to four degrees, and the impulse derived from the escape wheel, as illustrated at Fig. 76, will often fall below eight degrees. Such watches will have a poor motion and tick loud enough to keep a policeman awake. Trials with actual watches, with such a device as we have just described, in conjunction with a careful study of the acting parts, especially if aided by a large model, such as we have described, will soon bring the student to a degree of skill unknown to the old-style workman, who, if a poor escapement bothered him, would bend back the banking pins or widen the slot in the fork.

Fig. 78

We hold that educating our repair workmen up to a high knowledge of what is required to constitute a high-grade escapement, will have a beneficial effect on manufacturers. When we wish to apply our device to the measurement of the escapement of three-quarter-plate watches, we will require another index hand, with the grasping end bent downward, as shown at Fig. 77. The idea with this form of index hand is, the bent-down jaws B', Fig. 77, grasp the fork as close to the pallet staff as possible, making an allowance for the acting center by so placing the index arc that the hand A will read correctly on the index D. Suppose, for instance, we place the jaws B' inside the pallet staff, we then place the index arc so the hand reads to the arc indicated by the dotted arc m, Fig. 78, and if set outside of the pallet staff, read by the arc o.

HOW A BALANCE CONTROLS THE TIMEKEEPING OF A WATCH.

We think a majority of the fine lever escapements made abroad in this day have what is termed double-roller safety action. The chief gains to be derived from this form of safety action are: (1) Reducing the arc of fork and roller action; (2) reducing the friction of the guard point to a minimum. While it is entirely practicable to use a table roller for holding the jewel pin with a double-roller action, still a departure from that form is desirable, both for looks and because as much of the aggregate weight of a balance should be kept as far from the axis of rotation as possible.

We might as well consider here as elsewhere, the relation the balance bears to the train as a controlling power. Strictly speaking, *the balance and hairspring are the time measurers*, the train serving only two purposes: (*a*) To keep the balance in motion; (*b*) to classify and record the number of vibrations of the balance. Hence, it is of paramount importance that the vibrations of the balance should be as untrammeled as possible; this is why we urge reducing the arc of connection between the balance and fork to one as brief as is consistent with sound results. With a double-roller safety action we can easily reduce the fork action to eight degrees and the roller action to twenty-four degrees.

Inasmuch as satisfactory results in adjustment depend very much on the perfection of construction, we shall now dwell to some extent on the necessity of the several parts being made on correct principles. For instance, by reducing the arc of engagement between the fork and roller, we lessen the duration of any disturbing influence of escapement action.

To resume the explanation of why it is desirable to make the staff and all parts near the axis of the balance as light as possible, we would say it is the moving portion of the balance which controls the regularity of the intervals of vibration. To illustrate, suppose we have a balance only 3/8" in diameter, but of the same weight as one in an ordinary eighteen-size movement. We can readily see that such a balance would require but a very light hairspring to cause it to give the usual 18,000 vibrations to the hour. We can also understand, after a little thought, that such a balance would exert as much breaking force on its pivots as a balance of the same weight, but 3/4" in diameter acting against a very much stronger hairspring. There is another factor in the balance problem which deserves our attention, which factor is atmospheric resistance. This increases rapidly in proportion to the velocity.

HOW BAROMETRIC PRESSURE AFFECTS A WATCH.

The most careful investigators in horological mechanics have decided that a balance much above 75/100" in diameter, making 18,000 vibrations per hour, is not desirable, because of the varying atmospheric disturbances as indicated by barometric pressure. A balance with all of its weight as near the periphery as is consistent with strength, is what is to be desired for best results. It is the moving matter composing the balance, pitted against the elastic force of the hairspring, which we have to depend upon for the regularity of the timekeeping of a watch, and if we can take two grains' weight of matter from our roller table and place them in the rim or screws of the balance, so as to act to better advantage against the hairspring, we have disposed of these two grains so as to increase the efficiency of the controlling power and not increase the stress on the pivots.

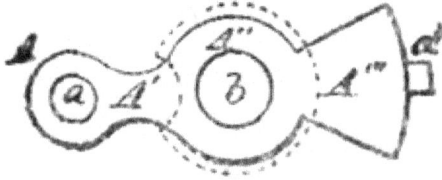

Fig. 79

We have deduced from the facts set forth, two axioms: (*a*) That we should keep the weight of our balance as much in the periphery as possible, consistent with due strength; (*b*) avoid excessive size from the disturbing effect of the air. We show at *A*, Fig. 79, the shape of the piece which carries the jewel pin. As shown, it consists of three parts: (1) The socket *A*, which receives the jewel pin *a*; (2) the part *A"* and hole *b*, which goes on the balance staff; (3) the counterpoise *A'''*, which makes up for the weight of the jewel socket *A*, neck *A'* and jewel pin. This counterpoise also makes up for the passing hollow *C* in the guard roller *B*, Fig. 80. As the piece *A* is always in the same relation to the roller *B*, the poise of the balance must always remain the same, no matter how the roller action is placed on the staff. We once saw a double roller of nearly the shape shown at Fig. 79, which had a small gold screw placed at *d*, evidently for the purpose of poising the double rollers; but, to our thinking, it was a sort of hairsplitting hardly worth the extra trouble. Rollers for very fine watches should be poised on the staff before the balance is placed upon it.

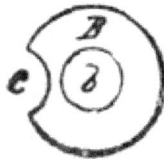

Fig. 80

We shall next give detailed instructions for drawing such a double roller as will be adapted for the large model previously described, which, as the reader will remember, was for ten degrees of roller action. We will also point out the necessary changes required to make it adapted for eight degrees of fork action. We would beg to urge again the advantages to be derived from constructing such a model, even for workmen who have had a long experience in escapements, our word for it they will discover a great many new wrinkles they never dreamed of previously.

It is important that every practical watchmaker should thoroughly master the theory of the lever escapement and be able to comprehend and understand at sight the faults and errors in such escapements, which, in the every-day practice of his profession, come to his notice. In no place is such knowledge more required than in fork and roller action. We are led to say the above chiefly for the benefit of a class of workmen who think there is a certain set of rules which, if they could be obtained, would enable them to set to rights any and all escapements. It is well to understand that no such system exists and that, practically, we must make one error balance another; and it is the "know how" to make such faults and errors counteract each other that enables one workman to earn more for himself or his employer in two days than another workman, who can file and drill as well as he can, will earn in a week.

PROPORTIONS OF THE DOUBLE-ROLLER ESCAPEMENT.

The proportion in size between the two rollers in a double-roller escapement is an open question, or, at least, makers seldom agree on it. Grossmann shows, in his work on the lever escapement, two sizes: (1) Half the diameter of the acting roller; (2) two-thirds of the size of the acting roller. The chief fault urged against a smaller safety roller is, that it necessitates longer horns to the fork to carry out the safety action. Longer horns mean more metal in the lever, and it is the conceded policy of all recent makers to have the fork and pallets as light as possible. Another fault pertaining to long horns is, when the horn does have to act as safety action, a greater friction ensues.

In all soundly-constructed lever escapements the safety action is only called into use in exceptional cases, and if the watch was lying still would theoretically never be required. Where fork and pallets are poised on their arbor, pocket motion (except torsional) should but very little affect the fork and pallet action of a watch, and torsional motion is something seldom brought to act on a watch to an extent to make it worthy of much consideration. In the double-roller action which we shall consider, we shall adopt three-fifths of the pitch diameter of the jewel-pin action as the proper size. Not but what the proportions given by Grossmann will do good service; but we adopt the proportions named because it enables us to use a light fork, and still the friction of the guard point on the roller is but little more than where a guard roller of half the diameter of the acting roller is employed.

The fork action we shall consider at present is ten degrees, but subsequently we shall consider a double-roller action in which the fork and pallet action is reduced to eight degrees. We shall conceive the play between the guard point and the safety roller as one degree, which will leave half a degree of lock remaining in action on the engaged pallet.

THEORETICAL ACTION OF DOUBLE ROLLER CONSIDERED.

In the drawing at Fig. 81 we show a diagram of the action of the double-roller escapement. The small circle at A represents the center of the pallet staff, and the one at B the center of the balance staff. The radial lines A d and A d' represent the arc of angular motion of fork action. The circle b b represents the pitch circle of the jewel pin, and the circle at c c the periphery of the guard or safety roller. The points established on the circle c c by intersection of the radial lines A d and A d' we will denominate the points h and h'. It is at these points the end of the guard point of the fork will terminate. In construction, or in delineating for construction, we show the guard enough short of the points h h' to allow the fork an angular motion of one degree, from A as a center, before said point would come in contact with the safety roller.

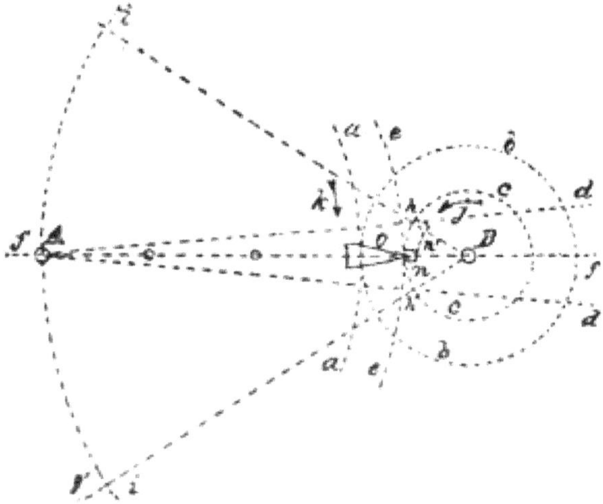

Fig. 81

We draw through the points h h', from B as a center, the radial lines B g and B g'. We measure this angle by sweeping the short arc i with any of the radii we have used for arc measurement in former delineations, and find it to be a trifle over sixty degrees. To give ourselves a practical object lesson, let us imagine that a real guard point rests on the circle c at h. Suppose we make a notch in the guard roller represented by the circle c, to admit such imaginary guard point, and then commence to revolve the circle c in the direction of the arrow j, letting the guard point rest constantly in such notch. When the notch n in c has been carried through thirty degrees of arc, counting from B as a center, the guard point, as relates to A as a center, would only have passed through an arc of five degrees. We show such a guard point and notch at o n. In fact, if a jewel pin was set to engage the fork on the pitch circle b a, the escapement would lock. To obviate such lock we widen the notch n to the extent indicated by the dotted lines n', allowing the guard point to fall back, so to speak, into the notch n, which really represents the passing hollow. It is not to be understood that the extended notch at n is correctly drawn as regards position, because when the guard point was on the line A f the point o would be in the center of the extended notch, or passing hollow. We shall next give the details of drawing the double roller, but before doing so we deemed it important to explain the action of such guard points more fully than has been done heretofore.

89

HOW TO DESIGN A DOUBLE-ROLLER ESCAPEMENT.

We have already given very desirable forms for the parts of a double-roller escapement, consequently we shall now deal chiefly with acting principles as regards the rollers, but will give, at Fig. 82, a very well proportioned and practical form of fork. The pitch circle of the jewel pin is indicated by the dotted circle a, and the jewel pin of the usual cylindrical form, with two-fifths cut away. The safety roller is three-fifths of the diameter of the pitch diameter of the jewel-pin action, as indicated by the dotted circle a.

The safety roller is shown in full outline at B', and the passing hollow at E. It will be seen that the arc of intersection embraced between the radial lines $B\ c$ and $B\ d$ is about sixty-one and a half degrees for the roller, but the angular extent of the passing hollow is only a little over thirty-two degrees. The passing hollow E is located and defined by drawing the radial line $B\ c$ from the center B through the intersection of radial line $A\ i$ with the dotted arc b, which represents the pitch circle of the safety roller. We will name this intersection the point l. Now the end of the guard point C terminates at the point l, and the passing hollow E extends on b sixteen degrees on each side of the radial line $B\ c$.

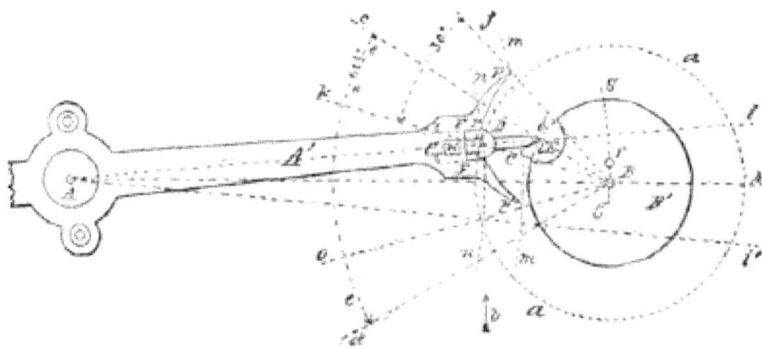

Fig. 82

The roller action is supposed to continue through thirty degrees of angular motion of the balance staff, and is embraced on the circle a between the radial line $B\ k$ and $B\ o$. To delineate the inner face of the horn p of the fork F we draw the short arc g, from A as a center, and on said arc locate at two degrees from the center at B the point f.

We will designate the upper angle of the outer face of the jewel pin D as the point s and, from A as a center, sweep through this point s the short arc $n\ n$. Parallel with the line $A\ i$ and at the distance of half the diameter of the jewel pin D, we draw the short lines $t\ t'$, which define the inner faces of the fork.

The intersection of the short line t with the arc n we will designate the point r. With our dividers set to embrace the space between the point r and the point f, we sweep the arc which defines the inner face of the prong of the fork. The space we just made use of is practically the same as the radius of the circle a, and consequently of the same curvature. Practically, the length of the guard point C' is made as long as will, with certainty, clear the safety roller B in all positions. While we set the point f at two degrees from the center B, still, in a well-constructed escapement, one and a half degrees should be sufficient, but the extra half degree will do no harm. If the roller B' is accurately made and the guard point C' properly fitted, the fork will not have half a degree of play.

The reader will remember that in the escapement model we described we cut down the drop to one degree, being less by half a degree than advised by Grossmann and Saunier. We also advised only one degree of lock. In the perfected lever escapement, which we shall describe and give working drawings for the construction of, we shall describe a detached lever escapement with only eight degrees fork and pallet action, with only three-fourths of a degree drop and three-fourths of a degree lock, which we can assure our readers is easily within the limits of practical construction by modern machinery.

HOW THE GUARD POINT IS MADE.

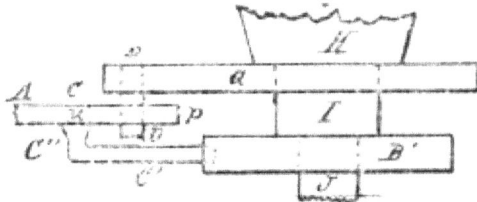

Fig. 83

91

The guard point C', as shown at Fig. 82, is of extremely simple construction. Back of the slot of the fork, which is three-fifths of the diameter of the jewel pin in depth, is made a square hole, as shown at u, and the back end of the guard point C is fitted to this hole so that it is rigid in position. This manner of fastening the guard point is equally efficient as that of attaching it with a screw, and much lighter—a matter of the highest importance in escapement construction, as we have already urged. About the best material for such guard points is either aluminum or phosphor bronze, as such material is lighter than gold and very rigid and strong. At Fig. 83 we show a side view of the essential parts depicted in Fig. 82, as if seen in the direction of the arrow v, but we have added the piece which holds the jewel pin D. A careful study of the cut shown at Fig. 82 will soon give the horological student an excellent idea of the double-roller action.

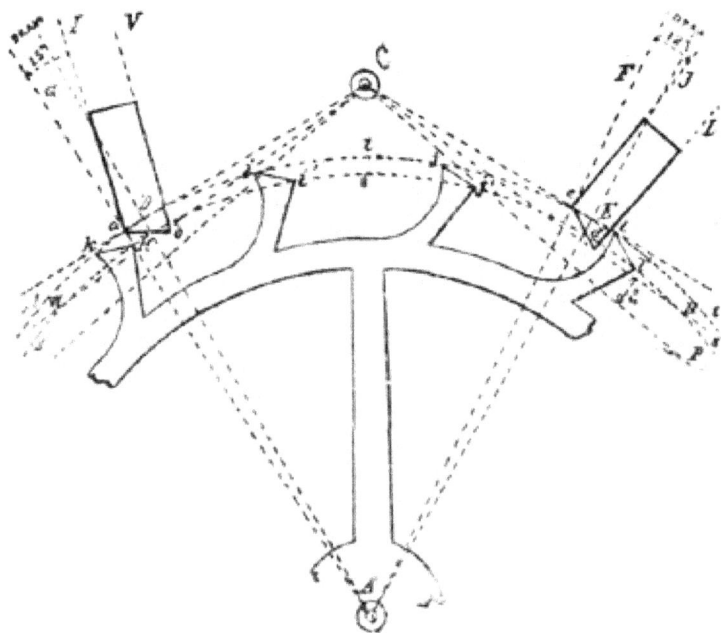

Fig. 84

We will now take up and consider at length why Saunier draws his entrance pallet with fifteen degrees draw and his exit pallet with only twelve degrees draw. To make ourselves more conversant

92

with Saunier's method of delineating the lever escapement, we reproduce the essential features of his drawing, Fig. 1, plate VIII, of his
"Modern Horology," in which he makes the draw of the locking face
of the entrance pallet fifteen degrees and his exit pallet twelve degrees.
In the cut shown at Fig. 84 we use the same letters of reference as he
employs. We do not quote his description or directions for delineation
because he refers to so much matter which he has previously given in
the book just referred to. Besides we cannot entirely endorse his methods of delineations for many reasons, one of which appears in the
drawing at Fig. 84.

MORE ABOUT TANGENTIAL LOCKINGS.

Most writers endorse the idea of tangential lockings, and
Saunier speaks of the escapement as shown at Fig. 84 as having such
tangential lockings, which is not the case. He defines the position of
the pallet staff from the circle t, which represents the extreme length of
the teeth; drawing the radial lines $A\ D$ and $A\ E$ to embrace an arc of
sixty degrees, and establishing the center of his pallet staff C at the
intersection of the lines $D\ C$ and $E\ C$, which are drawn at right angles
to the radial lines $A\ D$ and $A\ E$, and tangential to the circle t.

Here is an error; the lines defining the center of the pallet staff
should have been drawn tangent to the circle s, which represents the
locking angle of the teeth. This would have placed the center of the
pallet staff farther in, or closer to the wheel. Any person can see at a
glance that the pallets as delineated are not tangential in a true sense.
We have previously considered engaging friction and also repeatedly
have spoken of tangential lockings, but will repeat the idea of tangential lockings at Fig. 85. A tangential locking is neutral, or nearly so, as
regards engaging friction. For illustration we refer to Fig. 85, where A
represents the center of an escape wheel. We draw the radial lines $A\ y$
and $A\ z$ so that they embrace sixty degrees of the arcs s or t, which correspond to similar circles in Fig. 84, and represent the extreme extent
of the teeth and likewise the locking angle of such teeth. In fact, with
the club-tooth escapement all that part of a tooth which extends beyond the line s should be considered the same as the addendum in gear
wheels. Consequently, a tangential locking made to coincide with the

center of the impulse plane, as recommended by Saunier, would require the pallet staff to be located at C' instead of C, as he draws it. If
the angle k' of the tooth k in Fig. 84 was extended outward from the
center A so it would engage or rest on the locking face of the entrance
pallet as shown at Fig. 84, then the draw of the locking angle would
not be quite fifteen degrees; but it is evident no lock can take place
until the angle a of the entrance pallet has passed inside the circle s.
We would say here that we have added the letters s and t to the original drawings, as we have frequently to refer to these circles, and
without letters had no means of designation. Before the locking angle
k' of the tooth can engage the pallet, as shown in Fig. 84, the pallet
must turn on the center C through an angular movement of at least four
degrees. We show the situation in the diagram at Fig. 86, using the
same letters of reference for similar parts as in Fig. 84.

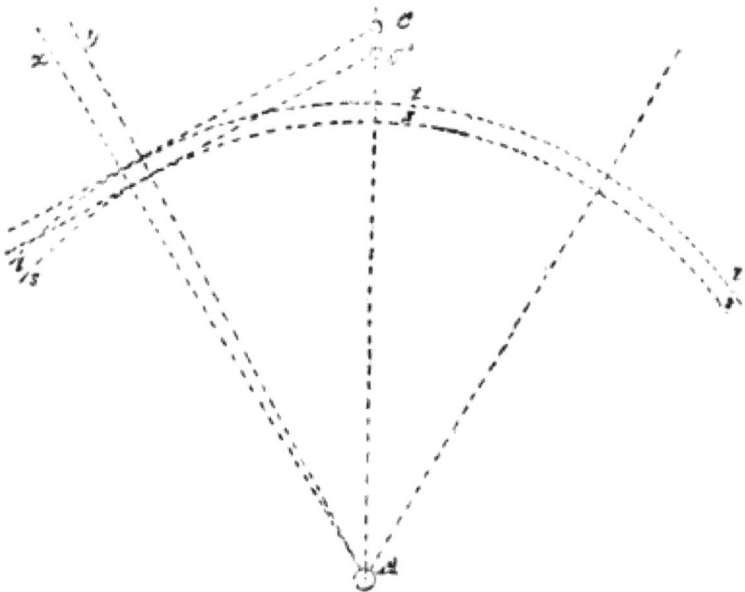

Fig. 85

94

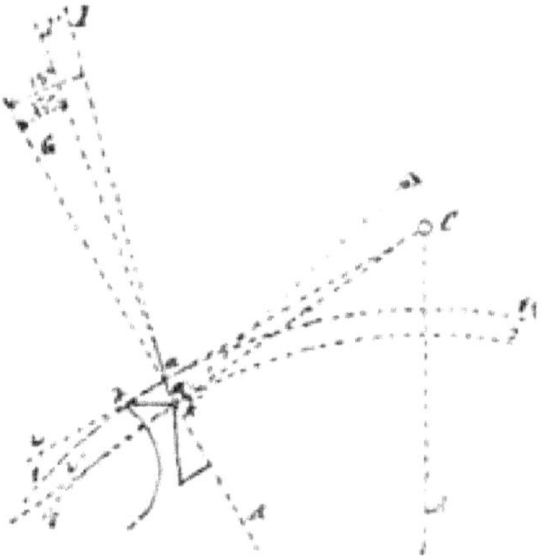

Fig. 86

As drawn in Fig. 84 the angle of draft *G a I* is equal to fifteen degrees, but when brought in a position to act as shown at *G a' I'*, Fig. 86, the draw is less even than twelve degrees. The angle *C a I* remains constant, as shown at *C a' I'*, but the relation to the radial *A G* changes when the pallet moves through the angle *w C w'*, as it must when locked. A tangential locking in the true sense of the meaning of the phrase is a locking set so that a pallet with its face coinciding with a radial line like *A G* would be neutral, and the thrust of the tooth would be tangent to the circle described by the locking angle of the tooth. Thus the center *C*, Fig. 86, is placed on the line *w'* which is tangent to the circle *s*; said line *w'* also being at right angles to the radial line *A G*.

The facts are, the problems relating to the club-tooth lever escapement are very intricate and require very careful analysis, and without such care the horological student can very readily be misled. Faulty drawings, when studying such problems, lead to no end of errors, and practical men who make imperfect drawings lead to the popular phrase, "Oh, such a matter may be all right in theory, but will not work in practice." We should always bear in mind that *theory, if right, must lead practice.*

CORRECT DRAWING REQUIRED.

If we delineate our entrance pallet to have a draw of twelve degrees when in actual contact with the tooth, and then construct in exact conformity with such drawings, we will find our lever to "hug the banks" in every instance. It is inattention to such details which produces the errors of makers complained of by Saunier in section 696 of his "Modern Horology," and which he attempts to correct by drawing the locking face at fifteen degrees draw.

We shall show that neither C nor C', Fig. 85, is the theoretically correct position for the pallet center for a tangential locking.

We will now take up the consideration of a club-tooth lever escapement with circular pallets and tangential lockings; but previous to making the drawings we must decide several points, among which are the thickness of the pallet arms, which establishes the angular motion of the escape wheel utilized by such pallet arms, and also the angular motion imparted to the pallets by the impulse faces of the teeth. We will, for the present, accept the thickness of the arms as being equivalent to five degrees of angular extent of the pitch circle of the escape wheel.

In making our drawings we commence, as on former occasions, by establishing the center of our escape wheel at A, Fig. 87, and sweeping the arc a a to represent the pitch circle of such wheel. Through the center A we draw the vertical line A B, which is supposed to also pass through the center of the pallet staff. The intersection of the line A B with the arc a we term the point d, and from this point we lay off on said arc a thirty degrees each side of said intersection, and thus establish the points c b. From A, through the point c, we draw the line A c c'. On the arc a a and two and a half degrees to the left of the point c we establish the point f, which space represents half of the thickness of the entrance pallet. From A we draw through the point f the line $A f f'$. From f, and at right angles to said line $A f$, we draw the line $f e$ until it crosses the line A B.

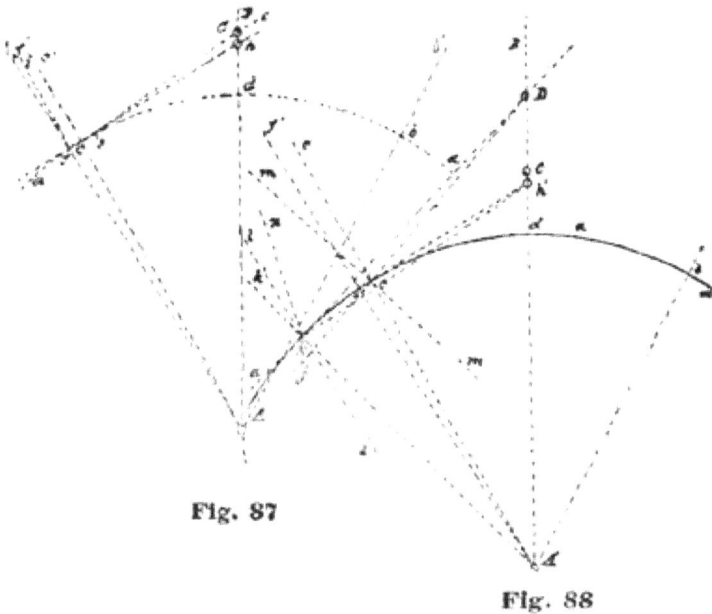

Fig. 87

Fig. 88

Now this line $f\,e$ is tangent to the arc a from the point f, and consequently a locking placed at the point f is a true tangential locking; and if the resting or locking face of a pallet was made to coincide with the line $A\,f$, such locking face would be strictly "dead" or neutral. The intersection of the line $f\,e$ with the line $A\,B$ we call the point C, and locate at this point the center of our pallet staff. According to the method of delineating the lever escapement by Moritz Grossmann the tangent line for locating the center of the pallet staff is drawn from the point c, which would locate the center of the pallet staff at the point h on the line $A\,B$.

Grossmann, in delineating his locking face for the draw, shows such face at an angle of twelve degrees to the radial line $A\,f$, when he should have drawn it twelve degrees to an imaginary line shown at $f\,i$, which is at right angles to the line $f\,h$. To the writer's mind this is not just as it should be, and may lead to misunderstanding and bad construction. We should always bear in mind the fact that the basis of a locking face is a neutral plane placed at right angles to the line of thrust, and the "draw" comes from a locking face placed at an angle to such neutral plane. A careful study of the diagram at Fig. 88 will give the reader correct ideas. If a tooth locks at the point c, the

97

tangential thrust would be on the line $c\ h'$, and a neutral locking face would be on the line $A\ c$.

NEUTRAL LOCKINGS.

To aid in explanation, let us remove the pallet center to D; then the line of thrust would be $c\ D$ and a neutral locking face would coincide with the line $m\ m$, which is at right angles to the line $c\ D$. If we should now make a locking face with a "draw" and at an angle to the line $c\ D$, say, for illustration, to correspond to the line $c\ c'$ (leaving the pallet center at D), we would have a strong draw and also a cruel engaging friction.

If, however, we removed the engaging tooth, which we have just conceived to be at c, to the point k on the arc $a'\ a'$, Fig. 88, the pallet center D would then represent a tangential locking, and a neutral pallet face would coincide with the radial line $A\ k'$; and a locking face with twelve degrees draw would coincide nearly with the line l. Let us next analyze what the effect would be if we changed the pallet center to h', Fig. 88, leaving the engaging tooth still at k. In this instance the line $l\ l$ would then coincide with a neutral locking face, and to obtain the proper draw we should delineate the locking face to correspond to the line $k\ n$, which we assume to be twelve degrees from $k\ l$.

It is not to be understood that we insist on precisely twelve degrees draw from a neutral plane for locking faces for lever pallets. What we do insist upon, however, is a "safe and sure draw" for a lever pallet which will hold a fork to the banks and will also return it to such banks if by accident the fork is moved away. We are well aware that it takes lots of patient, hard study to master the complications of the club-tooth lever escapement, but it is every watchmaker's duty to conquer the problem. The definition of "lock," in the detached lever escapement, is the stoppage or arrest of the escape wheel of a watch while the balance is left free or detached to perform the greater portion of its arc of vibration. "Draw" is a function of the locking parts to preserve the fork in the proper position to receive and act on the jewel pin of the balance.

It should be borne in mind in connection with "lock" and "draw," that the line of thrust as projected from the locked tooth of the

98

escape wheel should be as near tangential as practicable. This maxim applies particularly to the entrance pallet. We would beg to add that practically it will make but little odds whether we plant the center of our pallet staff at C or h, Fig. 87, provided we modify the locking and impulse angles of our pallets to conform to such pallet center. But it will not do to arrange the parts for one center and then change to another.

PRACTICAL HINTS FOR LEVER ESCAPEMENTS.

Apparently there seems to be a belief with very many watchmakers that there is a set of shorthand rules for setting an escapement, especially in American watches, which, if once acquired, conquers all imperfections. Now we wish to disabuse the minds of our readers of any such notions. Although the lever escapement, as adopted by our American factories, is constructed on certain "lines," still these lines are subject to modifications, such as may be demanded for certain defects of construction. If we could duplicate every part of a watch movement perfectly, then we could have certain rules to go by, and fixed templets could be used for setting pallet stones and correcting other escapement faults.

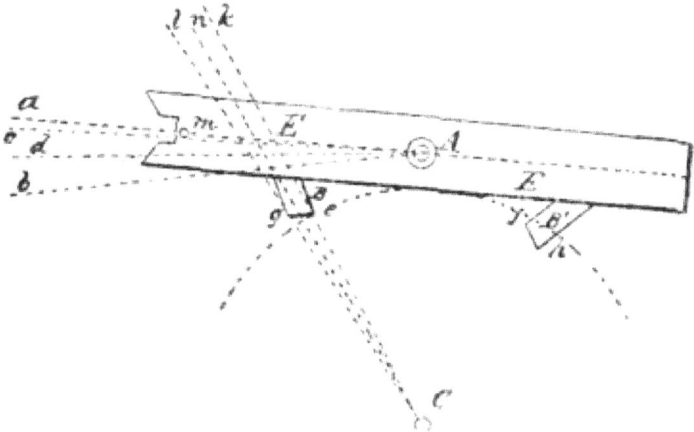

Fig. 89

Let us now make an analysis of the action of a lever escapement. We show at Fig. 89 an ordinary eighteen-size full-plate lever with fork and pallets. The dotted lines a b are supposed to represent an angular movement of ten degrees. Now, it is the function of the fork to carry the power of the train to the balance. How well the fork performs its office we will consider subsequently; for the present we are dealing with the power as conveyed to the fork by the pallets as shown at Fig. 89.

The angular motion between the lines a c (which represents the lock) is not only absolutely lost—wasted—but during this movement the train has to retrograde; that is, the dynamic force stored in the momentum of the balance has to actually turn the train backward and against the force of the mainspring. True, it is only through a very short arc, but the necessary force to effect this has to be discounted from the power stored in the balance from a former impulse. For this reason we should make the angular motion of unlocking as brief as possible. Grossmann, in his essay, endorses one and a half degrees as the proper lock.

In the description which we employed in describing the large model for illustrating the action of the detached lever escapement, we cut the lock to one degree, and in the description of the up-to-date lever escapement, which we shall hereafter give, we shall cut the lock down to three-quarters of a degree, a perfection easily to be attained by modern tools and appliances. We shall also cut the drop down to three-quarters of a degree. By these two economies we more than make up for the power lost in unlocking. With highly polished ruby or sapphire pallets ten degrees of draw is ample. But such draw must positively be ten degrees from a neutral locking face, not an escapement drawn on paper and called ten degrees, but when actually measured would only show eight and a half or nine degrees.

THE PERFECTED LEVER ESCAPEMENT.

With ten degrees angular motion of the lever and one and a half degrees lock, we should have eight and a half degrees impulse. The pith of the problem, as regards pallet action, for the practical workman can be embodied in the following question: What proportion of the power derived from the twelve degrees of angular motion of the escape wheel is really conveyed to the fork? The great leak of power as transmitted by the lever escapement to the balance is to be found in

100

the pallet action, and we shall devote special attention to finding and stopping such leaks.

WHEN POWER IS LOST IN THE LEVER ESCAPEMENT.

If we use a ratchet-tooth escape wheel we must allow at least one and a half degrees drop to free the back of the tooth; but with a club-tooth escape wheel made as can be constructed by proper skill and care, the drop can be cut down to three-quarters of a degree, or one-half of the loss with the ratchet tooth. We do not wish our readers to imagine that such a condition exists in most of the so-called fine watches, because if we take the trouble to measure the actual drop with one of the little instruments we have described, it will be found that the drop is seldom less than two, or even three degrees.

If we measure the angular movement of the fork while locked, it will seldom be found less than two or three degrees. Now, we can all understand that the friction of the locking surface has to be counted as well as the recoil of the draw. Locking friction is seldom looked after as carefully as the situation demands. Our factories make the impulse face of the pallets rounded, but leave the locking face flat. We are aware this condition is, in a degree, necessary from the use of exposed pallets. In many of the English lever watches with ratchet teeth, the locking faces are made cylindrical, but with such watches the pallet stones, as far as the writer has seen, are set "close"; that is, with steel pallet arms extending above and below the stone.

There is another feature of the club-tooth lever escapement that next demands our attention which we have never seen discussed. We refer to arranging and disposing of the impulse of the escape wheel to meet the resistance of the hairspring. Let us imagine the dotted line *A d*, Fig. 89, to represent the center of action of the fork. We can readily see that the fork in a state of rest would stand half way between the two banks from the action of the hairspring, and in the pallet action the force of the escape wheel, one tooth of which rests on the impulse face of a pallet, would be exerted against the elastic force of the hairspring. If the force of the mainspring, as represented by the escape-wheel tooth, is superior to the power of the hairspring, the watch starts itself. The phases of this important part of the detached lever escapement will be fully discussed.

ABOUT THE CLUB-TOOTH ESCAPEMENT.

We will now take up a study of the detached lever escapement as relates to pallet action, with the point specially in view of constructing an escapement which cannot "set" in the pocket, or, in other words, an escapement which will start after winding (if run down) without shaking or any force other than that supplied by the train as impelled by the mainspring. In the drawing at Fig. 90 we propose to utilize eleven degrees of escape-wheel action, against ten and a half, as laid down by Grossmann. Of this eleven degrees we propose to divide the impulse arc of the escape wheel in six and five degrees, six to be derived from the impulse face of the club tooth and five from the impulse plane of the pallet.

The pallet action we divide into five and four, with one degree of lock. Five degrees of pallet action is derived from the impulse face of the tooth and four from the impulse face of the pallet. The reader will please bear in mind that we do not give these proportions as imperative, because we propose to give the fullest evidence into the reader's hands and enable him to judge for himself, as we do not believe in laying down imperious laws that the reader must accept on our assertion as being correct. Our idea is rather to furnish the proper facts and put him in a situation to know for himself.

The reader is urged to make the drawings for himself on a large scale, say, an escape wheel 10" pitch diameter. Such drawings will enable him to realize small errors which have been tolerated too much in drawings of this kind. The drawings, as they appear in the cut, are one-fourth the size recommended, and many of the lines fail to show points we desire to call attention to. As for instance, the pallet center at B is tangential to the pitch circle a from the point of tooth contact at f. To establish this point we draw the radial lines $A\ c$ and $A\ d$ from the escape-wheel center A, as shown, by laying off thirty degrees on each side of the intersection of the vertical line i (passing through the centers $A\ B$) with the arc a, and then laying off two and a half degrees on a and establishing the point f, and through f from the center A draw the radial line $A\ f'$. Through the point f we draw the tangent line

102

b' b b'', and at the intersection of the line b with i we establish the center of our pallet staff at B. At two and a half degrees from the point c we lay off two and a half degrees to the right of said point and establish the point n, and draw the radial line A n n', which establishes the extent of the arc of angular motion of the escape wheel utilized by the pallet arm.

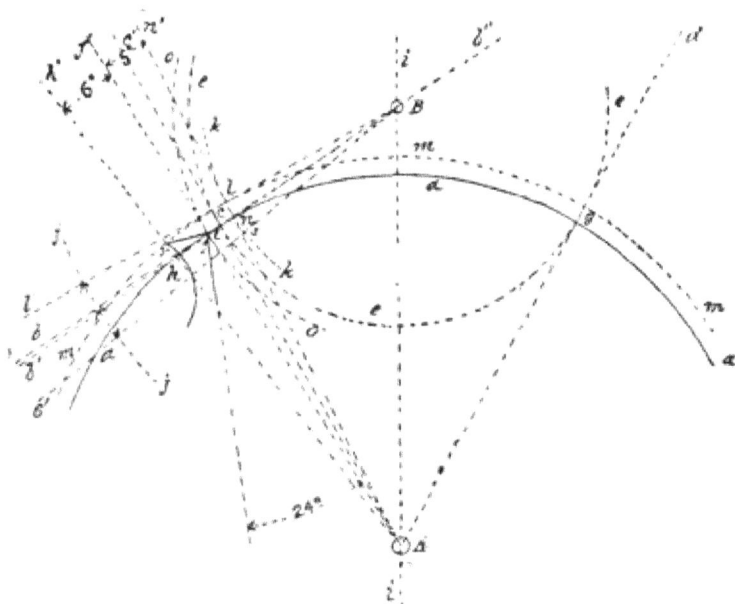

Fig. 90

We have now come to the point where we must exercise our reasoning powers a little. We know the locking angle of the escape-wheel tooth passes on the arc a, and if we utilize the impulse face of the tooth for five degrees of pallet or lever motion we must shape it to this end. We draw the short arc k through the point n, knowing that the inner angle of the pallet stone must rest on this arc wherever it is situated. As, for instance, when the locking face of the pallet is engaged, the inner angle of the pallet stone must rest somewhere on this arc (k) inside of a, and the extreme outer angle of the impulse face of the tooth must part with the pallet on this arc k.

103

HOW TO LOCATE THE PALLET ACTION.

With the parts related to each other as shown in the cut, to establish where the inner angle of the pallet stone is located in the drawing, we measure down on the arc k five degrees from its intersection with a, and establish the point s. The line $B\,b$, Fig. 90, as the reader will see, does not coincide with the intersection of the arcs a and k, and to conveniently get at the proper location for the inner angle of our pallet stone, we draw the line $B\,b'$, which passes through the point n located at the intersection of the arc a with the arc k. From B as a center we sweep the short arc j with any convenient radius of which we have a sixty-degree scale, and from the intersection of $B\,b'$ with j we lay off five degrees and draw the line $B\,s'$, which establishes the point s on the arc k. As stated above, we allow one degree for lock, which we establish on the arc o by laying off one degree on the arc j below its intersection with the line $B\,b$. We do not show this line in the drawing, from the fact that it comes so near to $B\,b'$ that it would confuse the reader. Above the arc a on the arc k at five degrees from the point n we establish the point l, by laying off five degrees on the arc j above the intersection of the line $B\,b$ with j.

The point l, Fig. 90, establishes where the outer angle of the tooth will pass the arc k to give five degrees of angular motion to the lever. From A as a center we sweep the arc m, passing through the point l. The intersection of the arc m with the line $A\,h$ we call the point r, and by drawing the right line $r\,f$ we delineate the impulse face of the tooth. On the arc o and one degree below its intersection with the line $B\,b$ we establish the point t, and by drawing a right line from t to s we delineate the impulse face of our entrance pallet.

"ACTION" DRAWINGS.

One great fault with most of our text books on horology lies in the fact that when dealing with the detached lever escapement the drawings show only the position of the pallets when locked, and many of the conditions assumed are arrived at by mental processes, without making the proper drawings to show the actual relation of the parts at the time such conditions exist. For illustration, it is often urged that there is a time in the action of the club-tooth lever escapement action when the incline on the tooth and the incline on the pallet present parallel surfaces, and consequently endure excessive friction, especially if the oil is a little thickened.

We propose to make drawings to show the exact position and relation of the entrance pallet and tooth at three intervals viz: (1) Locked; (2) the position of the parts when the lever has performed one-half of its angular motion; (3) when half of the impulse face of the tooth has passed the pallet. The position of the entrance pallet when locked is sufficiently well shown in Fig. 90 to give a correct idea of the relations with the entrance pallet; and to conform to statement (2), as above. We will now delineate the entrance pallet, not in actual contact, however, with the pallet, because if we did so the lines we employed would become confused. The methods we use are such that *we can delineate with absolute correctness either a pallet or tooth at any point in its angular motion.*

We have previously given instructions for drawing the pallet locked; and to delineate the pallet after five degrees of angular motion, we have only to conceive that we substitute the line s' for the line b'. All angular motions and measurements for pallet actions are from the center of the pallet staff at B. As we desire to now delineate the entrance pallet, it has passed through five degrees of angular motion and the inner angle s now lies on the pitch circle of the escape wheel, the angular space between the lines b' s' being five degrees, the line b''[**note: check this against the diagram-most other lines nave a two-letter identification] reducing the impulse face to four degrees.

DRAWING AN ESCAPEMENT TO SHOW ANGULAR MOTION.

To delineate our locking face we draw a line at right angles to the line B b'' from the point t, said point being located at the intersection of the arc o with the line B b''. To draw a line perpendicular to B b'' from the point t, we take a convenient space in our dividers and establish on the line B b'' the points x x' at equal distances from the point t. We open the dividers a little (no special distance) and sweep the short arcs x'' x''', as shown at Fig. 91. Through the intersection of the short arcs x'' x''' and to the point t we draw the line t y. The reader will see from our former explanations that the line t y represents the neutral plane of the locking face, and that to have the proper draw we must delineate the locking face of our pallet at twelve degrees. To do this we draw the line *t x'* at twelve degrees to the line *t y*, and proceed to outline our pallet faces as shown. We can now understand, after a moment's thought, that we can delineate the impulse face of a tooth at any point or place we choose by laying off six degrees on the arc *m*, and drawing radial lines from *A* to embrace such arc. To illustrate, suppose we draw the radial lines *w' w''* to embrace six degrees on the arc *a*. We make these lines contiguous to the entrance pallet *C* for convenience only. To delineate the impulse face of the tooth, we draw a line extending from the intersection of the radial line *A' w'* with the arc *m* to the intersection of the arc *a* with the radial line *A w''*.

Fig. 91

We next desire to know where contact will take place between the wheel-tooth *D* and pallet *C*. To determine this we sweep, with our dividers set so one leg rests at the escape-wheel center *A* and the other at the outer angle *t* of the entrance pallet, the short arc *t' w*. Where this arc intersects the line *w* (which represents the impulse face of the tooth) is where the outer angle *t* of the entrance pallet *C* will touch the impulse face of the tooth. To prove this we draw the radial line *A v* through the point where the short arc *t t'* passes through the impulse face *w* of the tooth *D*. Then we continue the line *w* to *n*, to represent the impulse face of the tooth, and then measure the angle *A w n* be-

106

tween the lines *w n* and *v A*, and find it to be approximately sixty-four degrees. We then, by a similar process, measure the angle *A t s'* and find it to be approximately sixty-six degrees. When contact ensues between the tooth *D* and pallet *C* the tooth *D* will attack the pallet at the point where the radial line *A v* crosses the tooth face. We have now explained how we can delineate a tooth or pallet at any point of its angular motion, and will next explain how to apply this knowledge in actual practice.

PRACTICAL PROBLEMS IN THE LEVER ESCAPEMENT.

To delineate our entrance pallet after one-half of the engaged tooth has passed the inner angle of the entrance pallet, we proceed, as in former illustrations, to establish the escape-wheel center at *A*, and from it sweep the arc *b*, to represent the pitch circle. We next sweep the short arcs *p s*, to represent the arcs through which the inner and outer angles of the entrance pallet move. Now, to comply with our statement as above, we must draw the tooth as if half of it has passed the arc *s*.

To do this we draw from *A* as a center the radial line *A j*, passing through the point *s*, said point *s* being located at the intersection of the arcs *s* and *b*. The tooth *D* is to be shown as if one half of it has passed the point *s*; and, consequently, if we lay off three degrees on each side of the point *s* and establish the points *d m*, we have located on the arc *b* the angular extent of the tooth to be drawn. To aid in our delineations we draw from the center *A* the radial lines *A d'* and *A m'*, passing through the points *d m*. The arc *a* is next drawn as in former instructions and establishes the length of the addendum of the escape-wheel teeth, the outer angle of our escape-wheel tooth being located at the intersection of the arc *a* with the radial line *A d'*.

As shown in Fig. 92, the impulse planes of the tooth *D* and pallet *C* are in contact and, consequently, in parallel planes, as mentioned on page 91. It is not an easy matter to determine at exactly what degree of angular motion of the escape wheel such condition takes place; because to determine such relation mathematically requires a knowledge of higher mathematics, which would require more study than most practical men would care to bestow, especially as they would have but very little use for such knowledge except for this problem and a few others in dealing with epicycloidal curves for the teeth of wheels.

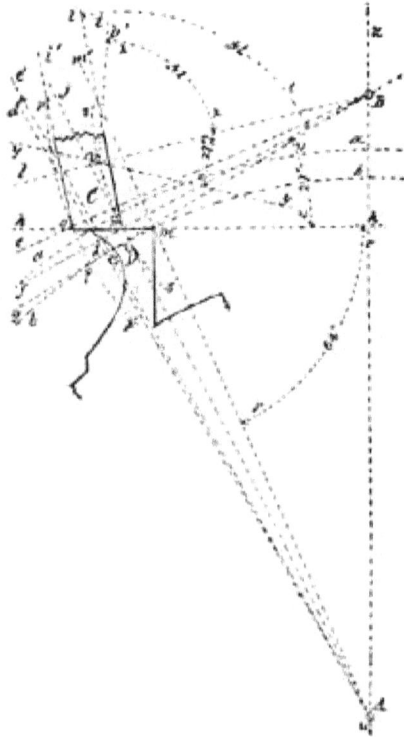

Fig. 92

For all practical purposes it will make no difference whether such parallelism takes place after eight or nine degrees of angular motion of the escape wheel subsequent to the locking action. The great point, as far as practical results go, is to determine if it takes place at or near the time the escape wheel meets the greatest resistance from the hairspring. We find by analysis of our drawing that parallelism takes place about the time when the tooth has three degrees of angular motion to make, and the pallet lacks about two degrees of angular movement for the tooth to escape. It is thus evident that the relations, as shown in our drawing, are in favor of the train or mainspring power over hairspring resistance as three is to two, while the average is only as eleven to ten; that is, the escape wheel in its entire effort passes

108

through eleven degrees of angular motion, while the pallets and fork move through ten degrees. The student will thus see we have arranged to give the train-power an advantage where it is most needed to overcome the opposing influence of the hairspring.

As regards the exalted adhesion of the parallel surfaces, we fancy there is more harm feared than really exists, because, to take the worst view of the situation, such parallelism only exists for the briefest duration, in a practical sense, because theoretically these surfaces never slide on each other as parallel planes. Mathematically considered, the theoretical plane represented by the impulse face of the tooth approaches parallelism with the plane represented by the impulse face of the pallet, arrives at parallelism and instantly passes away from such parallelism.

TO DRAW A PALLET IN ANY POSITION.

As delineated in Fig. 92, the impulse planes of the tooth and pallet are in contact; but we have it in our power to delineate the pallet at any point we choose between the arcs *p s*. To describe and illustrate the above remark, we say the lines *B e* and *B f* embrace five degrees of angular motion of the pallet. Now, the impulse plane of the pallet occupies four of these five degrees. We do not draw a radial line from *B* inside of the line *B e* to show where the outer angle of the impulse plane commences, but the reader will see that the impulse plane is drawn one degree on the arc *p* below the line *B e*. We continue the line *h h* to represent the impulse face of the tooth, and measure the angle *B n h* and find it to be twenty-seven degrees. Now suppose we wish to delineate the entrance pallet as if not in contact with the escape-wheel tooth—for illustration, say, we wish the inner angle of the pallet to be at the point *v* on the arc *s*. We draw the radial line *B l* through *v*; and if we draw another line so it passes through the point *v* at an angle of twenty-seven degrees to *B l*, and continue said line so it crosses the arc *p*, we delineate the impulse face of our pallet.

We measure the angle *i n B*, Fig. 92, and find it to be seventy-four degrees; we draw the line *v t* to the same angle with *v B*, and we define the inner face of our pallet in the new position. We draw a line parallel with *v t* from the intersection of the line *v y* with the arc *p*, and we define our locking face. If now we revolve the lines we have just drawn on the center *B* until the line *l B* coincides with the line *f B*, we will find the line *y y* to coincide with *h h*, and the line *v v'* with *n i*.

HIGHER MATHEMATICS APPLIED TO THE LEVER ESCAPEMENT.

We have now instructed the reader how to delineate either tooth or pallet in any conceivable position in which they can be related to each other. Probably nothing has afforded more efficient aid to practical mechanics than has been afforded by the graphic solution of abstruce mathematical problems; and if we add to this the means of correction by mathematical calculations which do not involve the highest mathematical acquirements, we have approached pretty close to the actual requirements of the practical watchmaker.

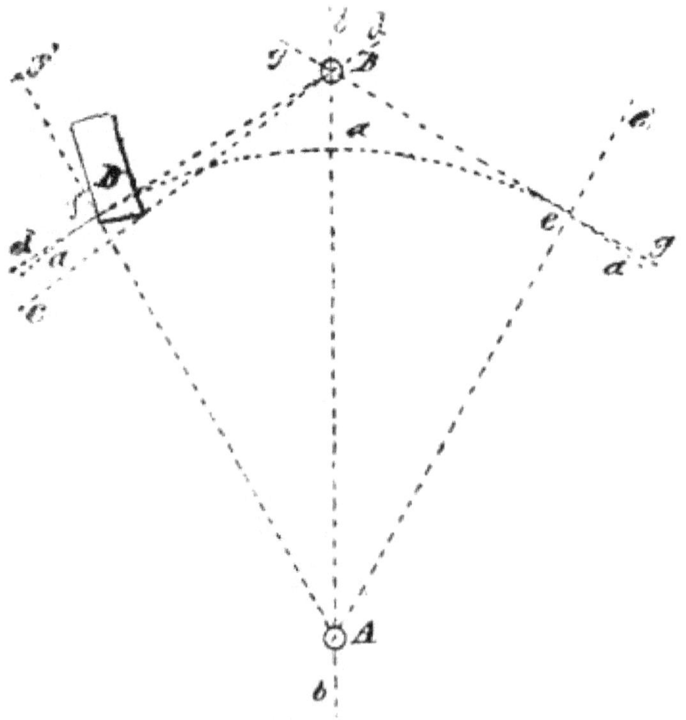

Fig. 93

To better explain what we mean, we refer the reader to Fig. 93, where we show preliminary drawings for delineating a lever escapement. We wish to ascertain by the graphic method the distance between the centers of action of the escape wheel and the pallet staff. We make our drawing very carefully to a given scale, as, for instance, the radius of the arc a is 5". After the drawing is in the condition shown at Fig. 93 we measure the distance on the line b between the points (centers) A B, and we thus by graphic means obtain a measure of the distance between A B. Now, by the use of trigonometry, we have the length of the line $A f$ (radius of the arc a) and all the angles given, to find the length of $f B$, or $A B$, or both $f B$ and $A B$. By adopting this policy we can verify the measurements taken from our drawings. Suppose we find by the graphic method that the distance between the points A B is 5.78", and by trigonometrical computation find the distance to be 5.7762". We know from this that there is .0038" to be accounted for somewhere; but for all practical purposes either measurement should be satisfactory, because our drawing is about thirty-eight times the actual size of the escape wheel of an eighteen-size movement.

HOW THE BASIS FOR CLOSE MEASUREMENTS IS OBTAINED.

Let us further suppose the diameter of our actual escape wheel to be .26", and we were constructing a watch after the lines of our drawing. By "lines," in this case, we mean in the same general form and ratio of parts; as, for illustration, if the distance from the intersection of the arc a with the line b to the point B was one-fifteenth of the diameter of the escape wheel, this ratio would hold good in the actual watch, that is, it would be the one-fifteenth part of .26". Again, suppose the diameter of the escape wheel in the large drawing is 10" and the distance between the centers A B is 5.78"; to obtain the actual distance for the watch with the escape wheel .26" diameter, we make a statement in proportion, thus: 10 : 5.78 :: .26 to the actual distance between the pivot holes of the watch. By computation we find the distance to be .15". These proportions will hold good in every part of actual construction.

All parts—thickness of the pallet stones, length of pallet arms, etc.—bear the same ratio of proportion. We measure the thickness of the entrance pallet stone on the large drawing and find it to be .47"; we make a similar statement to the one above, thus: 10 : .47 :: .26 to the actual thickness of the real pallet stone. By computation we find it to be .0122". All angular relations are alike, whether in the large drawing or the small pallets to match the actual escape wheel .26" in diameter. Thus, in the pallet *D*, Fig. 93, the impulse face, as reckoned from *B* as a center, would occupy four degrees.

MAKE A LARGE ESCAPEMENT MODEL.

Reason would suggest the idea of having the theoretical keep pace and touch with the practical. It has been a grave fault with many writers on horological matters that they did not make and measure the abstractions which they delineated on paper. We do not mean by this to endorse the cavil we so often hear—"Oh, that is all right in theory, but it will not work in practice." If theory is right, practice must conform to it. The trouble with many theories is, they do not contain all the elements or factors of the problem.

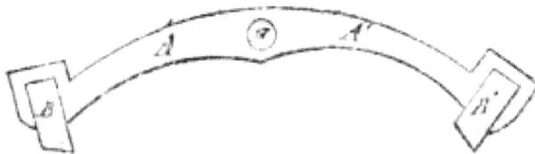

Fig. 94

Near the beginning of this treatise we advised our readers to make a large model, and described in detail the complete parts for such a model. What we propose now is to make adjustable the pallets and fork to such a model, in order that we can set them both right and wrong, and thus practically demonstrate a perfect action and also the various faults to which the lever escapement is subject. The pallet arms are shaped as shown at *A*, Fig. 94. The pallets *B B'* can be made of steel or stone, and for all practical purposes those made of steel answer quite as well, and have the advantage of being cheaper. A plate of sheet brass should be obtained, shaped as shown at *C*, Fig. 95. This plate is of thin brass, about No. 18, and on it are outlined the pallet arms shown at Fig. 94.

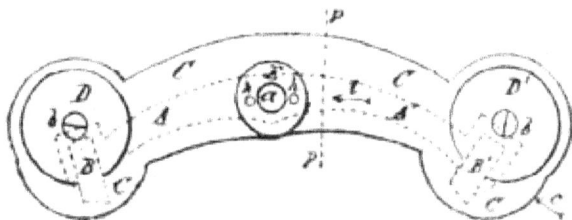

Fig. 95

Fig. 96

Fig. 97

To make the pallets adjustable, they are set in thick disks of sheet brass, as shown at D, Figs. 95, 96 and 97. At the center of the plate C is placed a brass disk E, Fig. 98, which serves to support the lever shown at Fig. 99. This disk E is permanently attached to the plate C. The lever shown at Fig. 99 is attached to the disk E by two screws, which pass through the holes h h. If we now place the brass pieces D D' on the plate C in such a way that the pallets set in them correspond exactly to the pallets as outlined on the plate C, we will find the action of the pallets to be precisely the same as if the pallet arms A A', Fig. 94, were employed.

Fig. 98

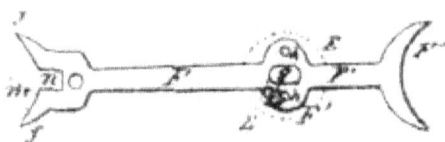

Fig. 99

To enable us to practically experiment with and to fully demonstrate all the problems of lock, draw, drop, etc., we make quite a large hole in C where the screws b come. To explain, if the screws b b were tapped directly into C, as they are shown at Fig. 95, we could only turn the disk D on the screw b; but if we enlarge the screw hole in C to three or four times the natural diameter, and then place the nut e under C to receive the screw b, we can then set the disks D D' and pallets B B' in almost any relation we choose to the escape wheel, and clamp the pallets fast and try the action. We show at Fig. 97 a view of the pallet B', disk D' and plate C (seen in the direction of the arrow c) as shown in Fig. 95.

PRACTICAL LESSONS WITH FORK AND PALLET ACTION.

It will be noticed in Fig. 99 that the hole g for the pallet staff in the lever is oblong; this is to allow the lever to be shifted back and forth as relates to roller and fork action. We will not bother about this now, and only call attention to the capabilities of such adjustments when required. At the outset we will conceive the fork F attached to the piece E by two screws passing through the holes h h, Fig. 99. Such an arrangement will insure the fork and roller action keeping right if they are put right at first. Fig. 100 will do much to aid in conveying a clear impression to the reader.

The idea of the adjustable features of our escapement model is to show the effects of setting the pallets wrong or having them of bad form. For illustration, we make use of a pallet with the angle too acute, as shown at B''', Fig. 101. The problem in hand is to find out by mechanical experiments and tests the consequences of such a change. It is evident that the angular motion of the pallet staff will be increased, and that we shall have to open one of the banking pins to allow the

114

engaging tooth to escape. To trace out *all* the consequences of this one little change would require a considerable amount of study, and many drawings would have to be made to illustrate the effects which would naturally follow only one such slight change.

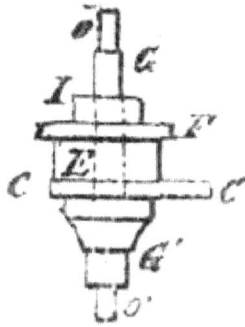

Fig. 100

Suppose, for illustration, we should make such a change in the pallet stone of the entrance pallet; we have increased the angle between the lines *k l* by (say) one and a half degrees; by so doing we would increase the lock on the exit pallet to three degrees, provided we were working on a basis of one and a half degrees lock; and if we pushed back the exit pallet so as to have the proper degree of lock (one and a half) on it, the tooth which would next engage the entrance pallet would not lock at all, but would strike the pallet on the impulse instead of on the locking face. Again, such a change might cause the jewel pin to strike the horn of the fork, as indicated at the dotted line *m*, Fig. 99.

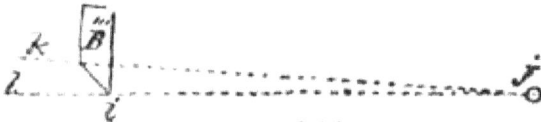

Fig. 101

Dealing with such and similar abstractions by mental process requires the closest kind of reasoning; and if we attempt to delineate all the complications which follow even such a small change, we will find the job a lengthy one. But with a large model having adjustable parts we provide ourselves with the means for the very best practical

115

solution, and the workman who makes and manipulates such a model will soon master the lever escapement.

QUIZ PROBLEMS IN THE DETACHED LEVER ESCAPEMENT.

Some years ago a young watchmaker friend of the writer made, at his suggestion, a model of the lever escapement similar to the one described, which he used to "play with," as he termed it—that is, he would set the fork and pallets (which were adjustable) in all sorts of ways, right ways and wrong ways, so he could watch the results. A favorite pastime was to set every part for the best results, which was determined by the arc of vibration of the balance. By this sort of training he soon reached that degree of proficiency where one could no more puzzle him with a bad lever escapement than you could spoil a meal for him by disarranging his knife, fork and spoon.

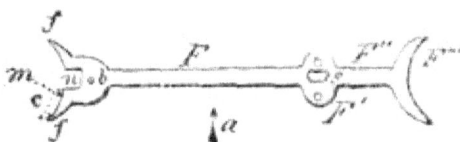

Fig. 102

Let us, as a practical example, take up the consideration of a short fork. To represent this in our model we take a lever as shown at Fig. 99, with the elongated slot for the pallet staff at *g*. To facilitate the description we reproduce at Fig. 102 the figure just mentioned, and also employ the same letters of reference. We fancy everybody who has any knowledge of the lever escapement has an idea of exactly what a "short fork" is, and at the same time it would perhaps puzzle them a good deal to explain the difference between a short fork and a roller too small.

Fig. 103

In our practical problems, as solved on a large escapement model, say we first fit our fork of the proper length, and then by the slot *g* move the lever back a little, leaving the bankings precisely as they were. What are the consequences of this slight change? One of the first results which would display itself would be discovered by the guard pin failing to perform its proper functions. For instance, the guard pin pushed inward against the roller would cause the engaged tooth to pass off the locking face of the pallet, and the fork, instead of returning against the banking, would cause the guard pin to "ride the roller" during the entire excursion of the jewel pin. This fault produces a scraping sound in a watch. Suppose we attempt to remedy the fault by bending forward the guard pin *b*, as indicated by the dotted outline *b'* in Fig. 103, said figure being a side view of Fig. 102 seen in the direction of the arrow *a*. This policy would prevent the engaged pallet from passing off of the locking face of the pallet, but would be followed by the jewel pin not passing fully into the fork, but striking the inside face of the prong of the fork at about the point indicated by the dotted line *m*. We can see that if the prong of the fork was extended to about the length indicated by the outline at *c*, the action would be as it should be.

To practically investigate this matter to the best advantage, we need some arrangement by which we can determine the angular motion of the lever and also of the roller and escape wheel. To do this, we provide ourselves with a device which has already been described, but of smaller size, for measuring fork and pallet action. The device to which we allude is shown at Figs. 104, 105 and 106. Fig. 104 shows only the index hand, which is made of steel about 1/20" thick and shaped as shown. The jaws *B″* are intended to grasp the pallet staff by the notches *e*, and hold by friction. The prongs *l l* are only to guard the staff so it will readily enter the notch *e*. The circle *d* is only to enable us to better hold the hand *B* flat.

117

Fig. 104

HOW TO MEASURE ESCAPEMENT ANGLES.

From the center of the notches *e* to the tip of the index hand *B'* the length is 2". This distance is also the radius of the index arc *C*. This index arc is divided into thirty degrees, with three or four supplementary degrees on each side, as shown. For measuring pallet action we only require ten degrees, and for roller action thirty degrees. The arc *C*, Fig. 105, can be made of brass and is about 1-1/2" long by 1/4" wide; said arc is mounted on a brass wire about 1/8" diameter, as shown at *k*, Fig. 106, which is a view of Fig. 105 seen in the direction of the arrow *i*. This wire *k* enters a base shown at *D E*, Fig. 106, which is provided with a set-screw at *j* for holding the index arc at the proper height to coincide with the hand *B*.

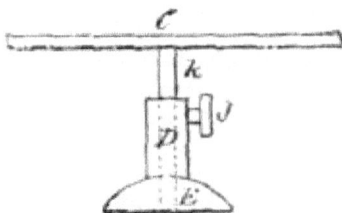

Fig. 105

A good way to get up the parts shown in Fig. 106 is to take a disk of thick sheet brass about 1" in diameter and insert in it a piece of brass wire about 1/4" diameter and 3/8" long, through which drill axially a hole to receive the wire *k*. After the jaws *B"* are clamped on the pallet staff, we set the index arc *C* so the hand *B'* will indicate the angular motion of the pallet staff. By placing the index hand *B* on the balance staff we can get at the exact angular duration of the engagement of the jewel pin in the fork.

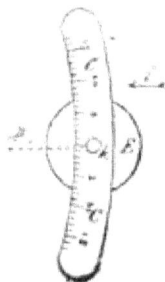

Fig. 106

Of course, it is understood that this instrument will also measure the angles of impulse and lock. Thus, suppose the entire angular
motion of the lever from bank to bank is ten degrees; to determine how
much of this is lock and how much impulse, we set the index arc C so
that the hand B' marks ten degrees for the entire motion of the fork,
and when the escapement is locked we move the fork from its bank
and notice by the arc C how many degrees the hand indicated before it
passed of its own accord to the opposite bank. If we have more than
one and a half degrees of lock we have too much and should seek to
remedy it. How? It is just the answers to such questions we propose to
give by the aid of our big model.

DETERMINATION OF "RIGHT" METHODS.

"Be sure you are right, then go ahead," was the advice of the
celebrated Davie Crockett. The only trouble in applying this motto to
watchmaking is to know when you are right. We have also often heard
the remark that there was only one right way, but any number of
wrong ways. Now we are inclined to think that most of the people who
hold to but one right way are chiefly those who believe all ways but
their own ways are wrong. Iron-bound rules are seldom sound even in
ethics, and are utterly impracticable in mechanics.

We have seen many workmen who had learned to draw a lever
escapement of a given type, and lived firm in the belief that all lever
escapements were wrong which were not made so as to conform to
this certain method. One workman believes in equidistant lockings,
another in circular pallets; each strong in the idea that their particular
and peculiar method of designing a lever escapement was the only one

119

to be tolerated. The writer is free to confess that he has seen lever escapements of both types, that is, circular pallets and equidistant lockings, which gave excellent results.

Another mooted point in the lever escapement is, to decide between the merits of the ratchet and the club-tooth escape wheel. English makers, as a rule, hold to the ratchet tooth, while Continental and American manufacturers favor the club tooth. The chief arguments in favor of the ratchet tooth are: (*a*) It will run without oiling the pallets; (*b*) in case the escape wheel is lost or broken it is more readily replaced, as all ratchet-tooth escape wheels are alike, either for circular pallets or equidistant lockings. The objections urged against it are: (*a*) Excessive drop; (*b*) the escape wheel, being frail, is liable to be injured by incompetent persons handling it; (*c*) this escapement in many instances does require to have the pallets oiled.

ESCAPEMENTS COMPARED.

(*a*) That a ratchet-tooth escape wheel requires more drop than a club tooth must be admitted without argument, as this form of tooth requires from one-half to three-fourths of a degree more drop than a club tooth; (*b*) as regards the frailty of the teeth we hold this as of small import, as any workman who is competent to repair watches would never injure the delicate teeth of an escape wheel; (*c*) ratchet-tooth lever escapements will occasionally need to have the pallets oiled. The writer is inclined to think that this defect could be remedied by proper care in selecting the stone (ruby or sapphire) and grinding the pallets in such a way that the escape-wheel teeth will not act against the foliations with which all crystalline stones are built up.

All workmen who have had an extended experience in repair work are well aware that there are some lever escapements in which the pallets absolutely require oil; others will seem to get along very nicely without. This applies also to American brass club-tooth escapements; hence, we have so much contention about oiling pallets. The writer does not claim to know positively that the pallet stones are at fault because some escapements need oiling, but the fact must admit of explanation some way, and is this not at least a rational solution? All persons who have paid attention to crystallography are aware that crystals are built up, and have lines of cleavage. In the manufacture of hole jewels, care must be taken to work with the axis of crystallization, or a smooth hole cannot be obtained.

The advantages claimed for the club-tooth escapement are many; among them may be cited (*a*) the fact that it utilizes a greater arc of impulse of the escape wheel; (*b*) the impulse being divided between the tooth and the pallet, permits greater power to be utilized at the close of the impulse. This feature we have already explained. It is no doubt true that it is more difficult to match a set of pallets with an escape wheel of the club-tooth order than with a ratchet tooth; still the writer thinks that this objection is of but little consequence where a workman knows exactly what to do and how to do it; in other words, is sure he is right, and can then go ahead intelligently.

It is claimed by some that all American escape wheels of a given grade are exact duplicates; but, as we have previously stated, this is not exactly the case, as they vary a trifle. So do the pallet jewels vary a little in thickness and in the angles. Suppose we put in a new escape wheel and find we have on the entrance pallet too much drop, that is, the tooth which engaged this pallet made a decided movement forward before the tooth which engaged the exit pallet encountered the locking face of said pallet. If we thoroughly understand the lever escapement we can see in an instant if putting in a thicker pallet stone for entrance pallet will remedy the defect. Here again we can study the effects of a change in our large model better than in an escapement no larger than is in an ordinary watch.

HOW TO SET PALLET STONES.

There have been many devices brought forward to aid the workman in adjusting the pallet stones to lever watches. Before going into the details of any such device we should thoroughly understand exactly what we desire to accomplish. In setting pallet stones we must take into consideration the relation of the roller and fork action. As has already been explained, the first thing to do is to set the roller and fork action as it should be, without regard in a great degree to pallet action.

Fig. 107

To explain, suppose we have a pallet stone to set in a full-plate movement. The first thing to do is to close the bankings so that the jewel pin will not pass out of the slot in the fork on either side; then gradually open the bankings until the jewel pin will pass out. This will be understood by inspecting Fig. 107, where *A A'* shows a lever fork as if in contact with both banks, and the jewel pin, represented at *B B''*, just passes the angle *a c'* of the fork. The circle described by the jewel pin *B* is indicated by the arc *e*. It is well to put a slight friction under the balance rim, in order that we can try the freedom of the guard pin. As a rule, all the guard pin needs is to be free and not touch the roller. The entire point, as far as setting the fork and bankings is concerned, is to have the fork and roller action sound. For all ordinary lever escapements the angular motion of the lever banked in as just described should be *about* ten degrees. As explained in former examples, if the fork action is entirely sound and the lever only vibrates through an arc of nine degrees, it is quite as well to make the pallets conform to this arc as to make the jewel pin carry the fork through full ten degrees. Again, if the lever vibrates through eleven degrees, it is as well to make the pallets conform to this arc.

The writer is well aware that many readers will cavil at this idea and insist that the workman should bring all the parts right on the basis of ten degrees fork and lever action. In reply we would say that no escapement is perfect, and it is the duty of the workman to get the best results he can for the money he gets for the job. In the instance

122

given above, of the escapement with nine degrees of lever action, when the fork worked all right, if we undertook to give the fork the ten degrees demanded by the stickler for accuracy we would have to set out the jewel pin or lengthen the fork, and to do either would require more time than it would to bring the pallets to conform to the fork and roller action. It is just this knowing how and the decision to act that makes the difference in the workman who is worth to his employer twelve or twenty-five dollars per week.

We have described instruments for measuring the angle of fork and pallet action, but after one has had experience he can judge pretty nearly and then it is seldom necessary to measure the angle of fork action as long as it is near the proper thing, and then bring the pallets to match the escape wheel after the fork and roller action is as it should be—that is, the jewel pin and fork work free, the guard pin has proper freedom, and the fork vibrates through an arc of about ten degrees.

Usually the workman can manipulate the pallets to match the escape wheel so that the teeth will have the proper lock and drop at the right instant, and again have the correct lock on the next succeeding pallet. The tooth should fall but a slight distance before the tooth next in action locks it, because all the angular motion the escape wheel makes except when in contact with the pallets is just so much lost power, which should go toward giving motion to the balance.

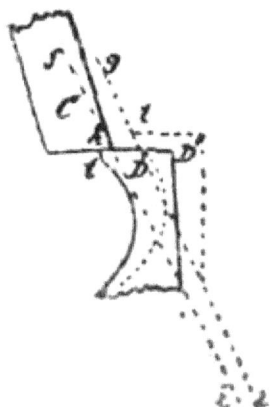

Fig. 108

123

There seems to be a little confusion in the use of the word "drop" in horological phrase, as it is used to express the act of parting of the tooth with the pallet. The idea will be seen by inspecting Fig. 108, where we show the tooth D and pallet C as about parting or dropping. When we speak of "banking up to the drop" we mean we set the banking screws so that the teeth will just escape from each pallet. By the term "fall" we mean the arc the tooth passes through before the next pallet is engaged. This action is also illustrated at Fig. 108, where the tooth D, after dropping from the pallet C, is arrested at the position shown by the dotted outline. We designate this arc by the term "fall," and we measure this motion by its angular extent, as shown by the dotted radial lines $i f$ and $i g$. As we have explained, this fall should only extend through an arc of one and a half degrees, but by close escapement matching this arc can be reduced to one degree, or even a trifle less.

We shall next describe an instrument for holding the escape wheel and pallets while adjusting them. As shown at Fig. 107, the fork A' is banked a little close and the jewel pin as shown would, in some portions, rub on C', making a scraping sound.

HOW TO MAKE AN ESCAPEMENT MATCHING TOOL.

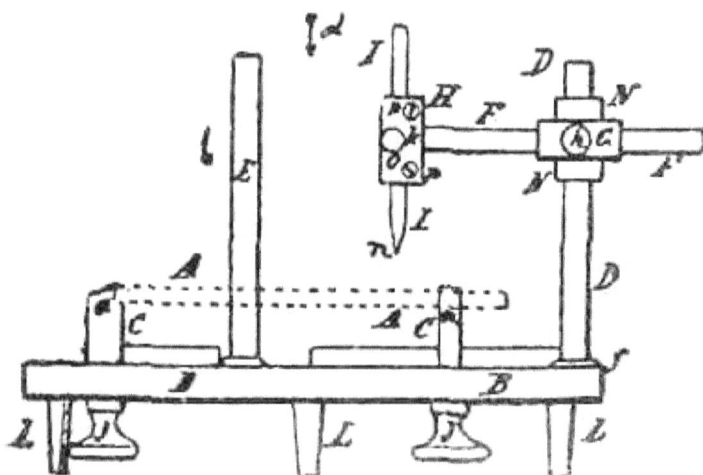

Fig. 109

A point has now been reached where we can use an escapement matcher to advantage. There are several good ones on the market, but we can make one very cheaply and also add our own improvements. In making one, the first thing to be provided is a movement holder. Any of the three-jaw types of such holders will answer, provided the jaws hold a movement plate perfectly parallel with the bed of the holder. This will be better understood by inspecting Fig. 109, which is a side view of a device of this kind seen edgewise in elevation. In this *B* represents the bed plate, which supports three swing jaws, shown at *C*, Figs. 109 and 110. The watch plate is indicated by the parallel dotted lines *A*, Fig. 109. The seat *a* of the swing jaws *C* must hold the watch plate *A* exactly parallel with the bed plate *B*. In the cheap movement holders these seats (*a*) are apt to be of irregular heights, and must be corrected for our purpose. We will take it for granted that all the seats *a* are of precisely the same height, measured from *B*, and that a watch plate placed in the jaws *C* will be held exactly parallel with the said bed *B*. We must next provide two pillars, shown at *D E*, Figs. 109 and 111. These pillars furnish support for sliding centers which hold the top pivots of the escape wheel and pallet staff while we are testing the depths and adjusting the pallet stones. It will be understood that these pillars *D E* are at right angles to the plane of the bed *B*, in order that the slides like *G N* on the pillars *D E* move exactly vertical. In fact, all the parts moving up and down should be accurately made, so as not to destroy the depths taken from the watch plate *A*. Suppose, to illustrate, that we place the plate *A* in position as shown, and insert the cone point *n*, Figs. 109 and 112, in the pivot hole for the pallet staff, adjusting the slide *G N* so that the cone point rests accurately in said pivot hole. It is further demanded that the parts *I H F G N D* be so constructed and adjusted that the sliding center *I* moves truly vertical, and that we can change ends with said center *I* and place the hollow cone end *m*, Fig. 112, so it will receive the top pivot of the pallet staff and hold it exactly upright.

125

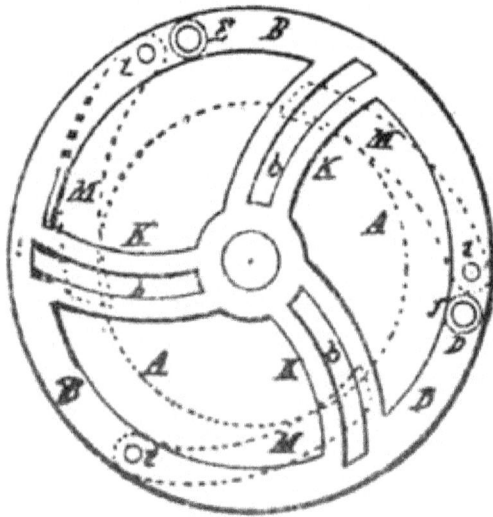

Fig. 110

Fig.111

Fig. 112

The idea of the sliding center *I* is to perfectly supply the place of the opposite plate of the watch and give us exactly the same practical depths as if the parts were in their place between the plates of the movement. The foot of the pillar *D* has a flange attached, as shown at *f*, which aids in holding it perfectly upright. It is well to cut a screw on

126

D at D', and screw the flange f on such screw and then turn the lower face of f flat to aid in having the pillar D perfectly upright.

DETAILS OF FITTING UP ESCAPEMENT MATCHER.

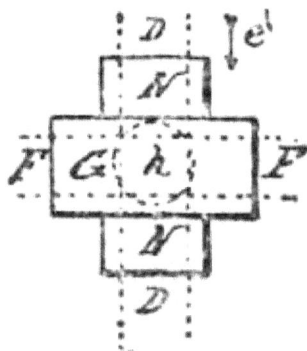

Fig. 113

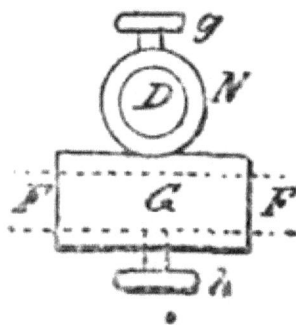

Fig. 114

It is well to fit the screw D' loosely, so that the flange f will come perfectly flat with the upper surface of the base plate B. The slide $G\,N$ on the pillar D can be made of two pieces of small brass tube, one fitting the pillar D and the other the bar F. The slide $G\,N$ is held in position by the set screw g, and the rod F by the set screw h.

127

The piece *H* can be permanently attached to the rod *F*. We show separate at Figs. 113 and 114 the slide *G N* on an enlarged scale from Fig. 109. Fig. 114 is a view of Fig. 113 seen in the direction of the arrow *e*. All joints and movable parts should work free, in order that the center *I* may be readily and accurately set. The parts *H F* are shown separate and enlarged at Figs. 115 and 116. The piece *H* can be made of thick sheet brass securely attached to *F* in such a way as to bring the V-shaped groove at right angles to the axis of the rod *F*. It is well to make the rod *F* about 1/8" in diameter, while the sliding center *I* need not be more than 1/16" in diameter. The cone point *n* should be hardened to a spring temper and turned to a true cone in an accurately running wire chuck.

Fig. 115

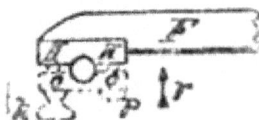

Fig. 116

The hollow cone end *m* of *I* should also be hardened, but this is best done after the hollow cone is turned in. The hardening of both ends should only be at the tips. The sliding center *I* can be held in the V-shaped groove by two light friction springs, as indicated at the dotted lines *s s*, Fig. 115, or a flat plate of No. 24 or 25 sheet brass of the size of *H* can be employed, as shown at Figs. 116 and 117, where *o* represents the plate of No. 24 brass, *p p* the small screws attaching the plate *o* to *H*, and *k* a clamping screw to fasten *I* in position. It will be found that the two light springs *s s*, Fig. 115 will be the most satisfactory. The wire legs, shown at *L*, will aid in making the device set steady. The pillar *E* is provided with the same slides and other parts as described and illustrated as attached to *D*. The position of the pillars *D* and *E* are indicated at Fig. 110.

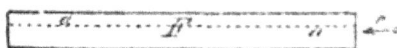

Fig. 118

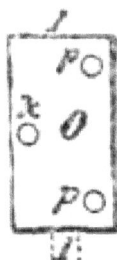

Fig. 117

Fig. 119

We will next tell how to flatten F to keep H exactly vertical. To aid in explanation, we will show (enlarged) at Fig. 118 the bar F shown in Fig. 109. In flattening such pieces to prevent turning, we should cut away about two-fifths, as shown at Fig. 119, which is an end view of Fig. 118 seen in the direction of the arrow c. In such flattening we should not only cut away two-fifths at one end, but we must preserve this proportion from end to end. To aid in this operation we make a fixed gage of sheet metal, shaped as shown at I, Fig. 120.

ESCAPEMENT MATCHING DEVICE DESCRIBED.

In practical construction we first file away about two-fifths of F and then grind the flat side on a glass slab to a flat, even surface and, of course, equal thickness from end to end. We reproduce the sleeve G as shown at Fig. 113 as if seen from the left and in the direction of the axis of the bar F. To prevent the bar F turning on its axis, we insert in the sleeve G a piece of wire of the same size as F but with three-fifths cut away, as shown at y, Fig. 121. This piece y is soldered in the sleeve

129

G so its flat face stands vertical. To give service and efficiency to the screw h, we thicken the side of the sleeve F by adding the stud w, through which the screw h works. A soft metal plug goes between the screw h and the bar F, to prevent F being cut up and marred. It will be seen that we can place the top plate of a full-plate movement in the device shown at Fig. 109 and set the vertical centers I so the cone points n will rest in the pivot holes of the escape wheel and pallets. It is to be understood that the lower side of the top plate is placed uppermost in the movement holder.

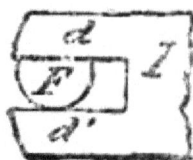

Fig. 120

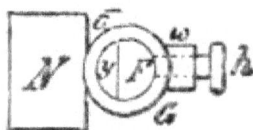

Fig. 121

If we now reverse the ends of the centers I and let the pivots of the escape wheel and pallet staff rest in the hollow cones of these centers I, we have the escape wheel and pallets in precisely the same position and relation to each other as if the lower plate was in position. It is further to be supposed that the balance is in place and the cock screwed down, although the presence of the balance is not absolutely necessary if the banking screws are set as directed, that is, so the jewel pin will just freely pass in and out of the fork.

HOW TO SET PALLET STONES.

We have now come to setting or manipulating the pallet stones so they will act in exact conjunction with the fork and roller. To do this we need to have the shellac which holds the pallet stones heated enough to make it plastic. The usual way is to heat a piece of metal and place it in close proximity to the pallets, or to heat a pair of pliers and clamp the pallet arms to soften the cement.

Of course, it is understood that the movement holder cannot be moved about while the stones are being manipulated. The better way is to set the movement holder on a rather heavy plate of glass or metal, so that the holder will not jostle about; then set the lamp so it will do its duty, and after a little practice the setting of a pair of pallet stones to perfectly perform their functions will take but a few minutes. In fact, if the stones will answer at all, three to five minutes is as much time as one could well devote to the adjustment. The reader will see that if the lever is properly banked all he has to do is to set the stones so the lock, draw and drop are right, when the entire escapement is as it should be, and will need no further trial or manipulating.

II. THE CYLINDER ESCAPEMENT

There is always in mechanical matters an underlying combination of principles and relations of parts known as "theory." We often hear the remark made that such a thing may be all right in theory, but will not work in practice. This statement has no foundation in fact. If a given mechanical device accords strictly with theory, it will come out all right practically. *Mental conceptions* of a machine are what we may term their theoretical existence.

When we make drawings of a machine mentally conceived, we commence its mechanical construction, and if we make such drawings to scale, and add a specification stating the materials to be employed, we leave only the merest mechanical details to be carried out; the brain work is done and only finger work remains to be executed.

With these preliminary remarks we will take up the consideration of the cylinder escapement invented by Robert Graham about the year 1720. It is one of the two so-called frictional rest dead-beat escapements which have come into popular use, the other being the duplex. Usage, or, to put it in other words, experience derived from the actual manufacture of the cylinder escapement, settled the best forms and proportions of the several parts years ago. Still, makers vary slightly on certain lines, which are important for a man who repairs such watches to know and be able to carry out, in order to put them in a condition to perform as intended by the manufacturers. It is not knowing these lines which leaves the average watchmaker so much at sea. He cuts and moves and shifts parts about to see if dumb luck will not supply the correction he does not know how to make. This requisite knowledge does not consist so much in knowing how to file or grind as it does in discriminating where such application of manual dexterity is to be applied. And right here let us make a remark to which we will call attention again later on. The point of this remark lies in the question—How many of the so-called practical watchmakers could tell you what proportion of a cylinder should be cut away from the half shell? How many could explain the difference between the "real" and "apparent" lift? Comparatively few, and yet a knowledge of these things is as important for a watchmaker as it is for a surgeon to understand the action of a man's heart or the relations of the muscles to the bones.

ESSENTIAL PARTS OF THE CYLINDER ESCAPEMENT.

The cylinder escapement is made up of two essential parts, viz.: the escape wheel and the cylinder. The cylinder escape wheel in all modern watches has fifteen teeth, although Saunier, in his "Modern Horology," delineates a twelve-tooth wheel for apparently no better reason than because it was more easily drawn. We, in this treatise, will consider both the theoretical action and the practical construction, but more particularly the repair of this escapement in a thorough and complete manner.

At starting out, we will first agree on the names of the several parts of this escapement, and to aid us in this we will refer to the accompanying drawings, in which Fig. 122 is a side elevation of a cylinder complete and ready to have a balance staked on to it. Fig. 123 shows the cylinder removed from the balance collet. Figs. 124 and 125 show the upper and lower plugs removed from the cylinder. Fig. 126 is a horizontal section of Fig. 122 on the line *i*. Fig. 127 is a side view of one tooth of a cylinder escape wheel as if seen in the direction of the arrow *f* in Fig. 126. Fig. 128 is a top view of two teeth of a cylinder escape wheel. The names of the several parts usually employed are as follows:

A.	—Upper or Main Shell.
A'.	—Half Shell.
A".	—Column.
A'".	—Small Shell.
B B' B".	—Balance Collet.
G.	—Upper Plug.
H.	—Lower Plug.
g.	—Entrance Lip of Cylinder.
h.	—Exit Lip of Cylinder.
c.	—Banking Slot.
C.	—Tooth.
D.	—U arm.
E.	—Stalk of Pillar.
I.	—U space
l.	—Point of Tooth.
k.	—Heel of Tooth.

The cylinder escapement has two engagements or actions, during the passage of each tooth; that is, one on the outside of the cylinder and one on the inside of the shell. As we shall show later on, the cylinder escapement is the only positively dead-beat escapement in use, all others, even the duplex, having a slight recoil during the process of escaping.

When the tooth of a cylinder escape wheel while performing its functions, strikes the cylinder shell, it rests dead on the outer or inner surface of the half shell until the action of the balance spring has brought the lip of the cylinder so that the impulse face of the tooth commences to impart motion or power to the balance.

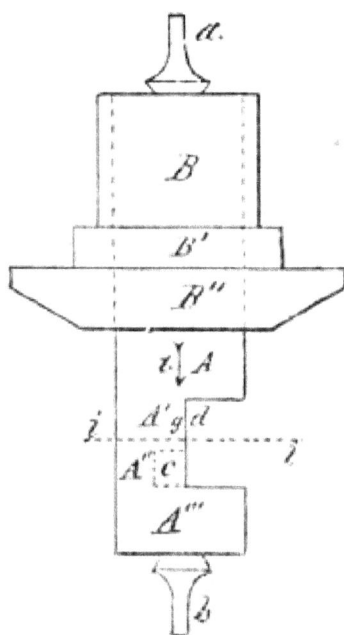

Fig. 122

Fig. 123

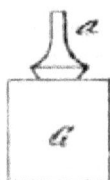

Fig. 124

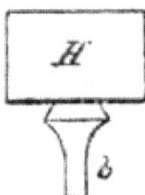

Fig. 125

Fig. 126

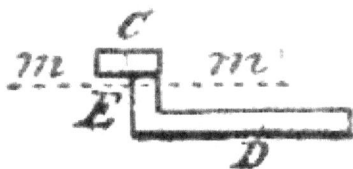

Fig. 127

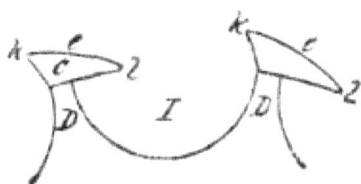

Fig. 128

Most writers on horological matters term this act the "lift," which name was no doubt acquired when escapements were chiefly confined to pendulum clocks. Very little thought on the matter will show any person who inspects Fig. 126 that if the tooth *C* is released or escapes from the inside of the half shell of the cylinder *A*, said cylinder must turn or revolve a little in the direction of the arrow *j*, and also that the next succeeding tooth of the escape wheel will engage the cylinder on the outside of the half shell, falling on the dead or neutral portion of said cylinder, to rest until the hairspring causes the cylinder to turn in the opposite direction and permitting the tooth now resting on the outside of the cylinder to assume the position shown on the drawing.

The first problem in our consideration of the theoretical action of the cylinder escapement, is to arrange the parts we have described so as to have these two movements of the escape wheel of like angular values. To explain what we mean by this, we must premise by saying, that as our escape wheel has fifteen teeth and we make each tooth give two impulses in alternate directions we must arrange to have these half-tooth movements exactly alike, or, as stated above, of equal angular values; and also each impulse must convey the same power or force to the balance. All escape wheels of fifteen teeth acting by half impulses must impel the balance during twelve degrees (minus the drop) of escape-wheel action; or, in other words, when a tooth passes out of the cylinder from the position shown at Fig. 126, the form of the impulse face of the tooth and the shape of the exit lip of the cylinder must be such during twelve degrees (less the drop) of the angular motion of the escape wheel. The entire power of such an escape wheel is devoted to giving impulse to the balance.

The extent of angular motion of the balance during such impulse is, as previously stated, termed the "lifting angle." This "lifting angle" is by horological writers again divided into real and apparent lifts. This last division is only an imaginary one, as the real lift is the one to be studied and expresses the arc through which the impulse face of the tooth impels the balance during the act of escaping, and so, as we shall subsequently show, should no more be counted than in the detached lever escapement, where a precisely similar condition exists, but is never considered or discussed.

We shall for the present take no note of this lifting angle, but confine ourselves to the problem just named, of so arranging and designing our escape-wheel teeth and cylinder that each half of the tooth space shall give equal impulses to the balance with the minimum of drop. To do this we will make a careful drawing of an escape-wheel tooth and cylinder on an enlarged scale; our method of making such drawings will be on a new and original system, which is very simple yet complete.

DRAWING THE CYLINDER ESCAPEMENT.

All horological—and for that matter all mechanical— drawings are based on two systems of measurements: (1) Linear extent; (2) angular movement. For the first measurement we adopt the inch and its decimals; for the second we adopt degrees, minutes and seconds. For measuring the latter the usual plan is to employ a protractor, which serves the double purpose of enabling us to lay off and delineate any angle and also to measure any angle obtained by the graphic method, and it is thus by this graphic method we propose to solve very simply some of the most abstruce problems in horological delineations. As an instance, we propose to draw our cylinder escapement with no other instruments than a steel straight-edge, showing one-hundredths of an inch, and a pair of dividers; the degree measurement being obtained from arcs of sixty degrees of radii, as will be explained further on.

In describing the method for drawing the cylinder escapement we shall make a radical departure from the systems usually laid down in text-books, and seek to simplify the formulas which have heretofore been given for such delineations. In considering the cylinder escapement we shall pursue an analytical course and strive to build up from the underlying principles. In the drawings for this purpose we shall commence with one having an escape wheel of 10" radius, and our first effort will be the primary drawing shown at Fig. 129. Here we establish the point A for the center of our escape wheel, and from this center sweep the short arc a a with a 10" radius, to represent the circumference of our escape wheel. From A we draw the vertical line A B, and from the intersection of said line with the arc a a we lay off twelve degree spaces on each side of the line A B on said arc a and establish the points b c. From A as a center we draw through the points b c the radial lines b' c'.

To define the face of the incline to the teeth we set our dividers to the radius of any of the convenient arcs of sixty degrees which we have provided, and sweep the arc t t. From the intersection of said

arc with the line *A b'* we lay off on said arc sixty-four degrees and establish the point *g* and draw the line *b g*. Why we take sixty-four
degrees for the angle *A b g* will be explained later on, when we are
discussing the angular motion of the cylinder. By dividing the eleventh
degree from the point *b* on the arc *a a* into thirds and taking two of
them, we establish the point *y* and draw the radial line *A y'*. Where this
line *A y'* intersects the line *b g* we name the point *n*, and in it is located
the point of the escape-wheel tooth. That portion of the line *b g* which
lies between the points *b* and *n* represents the measure of the inner diameter of the cylinder, and also the length of the chord of the arc
which rounds the impulse face of the tooth. We divide the space *b n*
into two equal portions and establish the point *e*, which locates the
position of the center of the cylinder. From *A* as a center and through
the point *e* we sweep the arc *e' e'*, and it is on this line that the points
establishing the center of the cylinder will in every instance be located.
From *A* as a center, through the point *n* we sweep the arc *k*, and on this
line we locate the points of the escape-wheel teeth. For delineating the
curved impulse faces of the escape-wheel teeth we draw from the point
e and at right angles to the line *b g* the line *e o*. We next take in our
dividers the radius of the arc *k*, and setting one leg at either of the
points *b* or *n*, establish with the other leg the point *p'* on the line *e o*,
and from the point *p'* as a center we sweep the arc *b v n*, which defines
the curve of the impulse faces of the teeth. From *A* as a center through
the point *p'* we sweep the arc *p*, and in all instances where we desire to
delineate the curved face of a tooth we locate either the position of the
point or the heel of such tooth, and setting one leg of our dividers at
such point, the other leg resting on the arc *p*, we establish the center
from which to sweep the arc defining the face of said tooth.

ADVANTAGES GAINED IN SHAPING.

The reason for giving a curved form to the impulse face of the
teeth of cylinder escape wheels are somewhat intricate, and the problem involves several factors. That there are advantages in so shaping
the incline or impulse face is conceded, we believe, by all recent
manufacturers. The chief benefit derived from such curved impulse
faces will be evident after a little thought and study of the situation and
relation of parts as shown in Fig. 129. It will be seen on inspection that
the angular motion imparted to the cylinder by the impulse face of the
tooth when curved as shown, is greater during the first half of the

twelve degrees of escape-wheel action than during the last half, thus giving the escape wheel the advantage at the time the balance spring increases its resistance to the passage of the escape-wheel tooth across the lip of the cylinder. Or, in other words, as the ratio of resistance of the balance spring increases, in a like ratio the curved form of the impulse face of the tooth gives greater power to the escape-wheel action in proportion to the angular motion of the escape wheel. Hence, in actual service it is found that cylinder watches with curved impulse planes to the escape-wheel teeth are less liable to set in the pocket than the teeth having straight impulse faces.

THE OUTER DIAMETER OF THE CYLINDER.

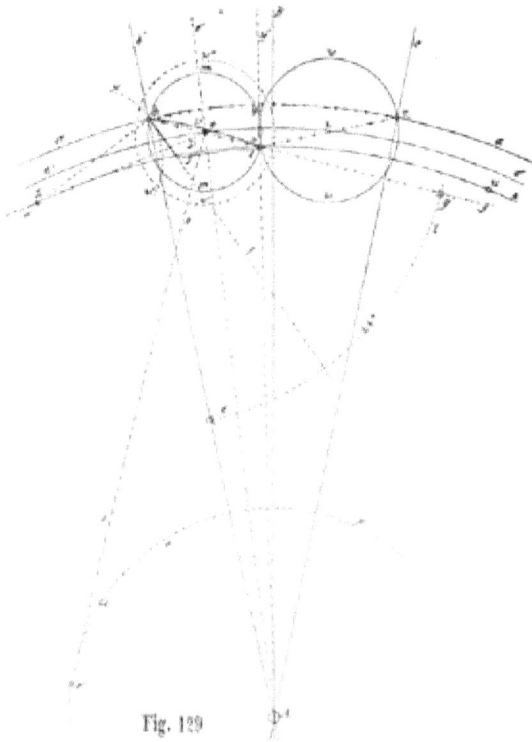

Fig. 129

To define the remainder of the form of our escape-wheel tooth we will next delineate the heel. To do this we first define the outer diameter of our cylinder, which is the extent from the point *n* to *c*, and

141

after drawing the line n c we halve the space and establish the point x, from which point as a center we sweep the circle w w, which defines the outer circumference of our cylinder. With our dividers set to embrace the extent from the point n to the point c we set one leg at the point b, and with the other leg establish on the arc k the point h. We next draw the line b h, and from the point b draw the line b f at right angle to the line b h. Our object for drawing these lines is to define the heel of our escape-wheel tooth by a right angle line tangent to the circle w, from the point b; which circle w represents the curve of the outer circumference of the cylinder. We shape the point of the tooth as shown to give it the proper stability, and draw the full line j to a curve from the center A. We have now defined the form of the upper face of the tooth. How to delineate the U arms will be taken up later on, as, in the present case, the necessary lines would confuse our drawing.

We would here take the opportunity to say that there is a great latitude taken by makers as regards the extent of angular impulse given to the cylinder, or, as it is termed, the "actual lift." This latitude governs to a great extent the angle A b g, which we gave as sixty-four degrees in our drawing. It is well to understand that the use of sixty-four degrees is based on no hard-and-fast rules, but varies back and forth, according as a greater or lesser angle of impulse or lift is employed.

In practical workshop usage the impulse angle is probably more easily estimated by the ratio between the diameter of the cylinder and the measured (by lineal measure) height of the impulse plane. Or, to be more explicit, we measure the radial extent from the center A between the arcs a k on the line A b, and use this for comparison with the outer diameter of the cylinder.

We can readily see that as we increase the height of the heel of the impulse face of our tooth we must also increase the angle of impulse imparted to the cylinder. With the advantages of accurate micrometer calipers now possessed by the horological student it is an easy matter to get at the angular extent of the real lift of any cylinder. The advantage of such measuring instruments is also made manifest in determining when the proper proportion of the cylinder is cut away for the half shell.

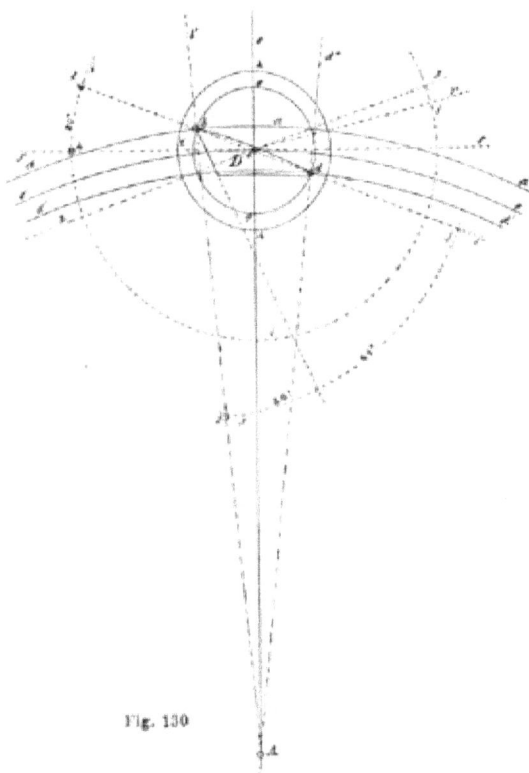

Fig. 130

In the older methods of watchmaking it was a very common rule to say, let the height of the incline of the tooth be one-seventh of the outer diameter of the cylinder, and at the same time the trade was furnished with no tools except a clumsy douzieme gage; but with micrometer calipers which read to one-thousandths of an inch such rules can be definitely carried into effect and not left to guess work. Let us compare the old method with the new: Suppose we have a new cylinder to put in; we have the old escape wheel, but the former cylinder is gone. The old-style workman would take a round broach and calculate the size of the cylinder by finding a place where the broach would just go between the teeth, and the size of the broach at this point was supposed to be the outer diameter of the cylinder. By our method we measure the diameter of the escape wheel in thousandths of an inch, and from this size calculate exactly what the diameter of the new cylinder should be in thousandths of an inch. Suppose, to further carry out our comparison, the escape wheel which is in the watch has teeth

143

which have been stoned off to permit the use of a cylinder which was too small inside, or, in fact, of a cylinder too small for the watch: in this case the broach system would only add to the trouble and give us a cylinder which would permit too much inside drop.

DRAWING A CYLINDER.

We have already instructed the pupil how to delineate a cylinder escape wheel tooth and we will next describe how to draw a cylinder. As already stated, the center of the cylinder is placed to coincide with the center of the chord of the arc which defines the impulse face of the tooth. Consequently, if we design a cylinder escape wheel tooth as previously described, and setting one leg of our compasses at the point e which is situated at the center of the chord of the arc which defines the impulse face of the tooth and through the points d and b we define the inside of our cylinder. We next divide the chord $d\ b$ into eight parts and set our dividers to five of these parts, and from e as a center sweep the circle h and define the outside of our cylinder. From A as a center we draw the radial line $A\ e'$. At right angles to the line A e' and through the point e we draw the line from e as a center, and with our dividers set to the radius of any of the convenient arcs which we have divided into sixty degrees, we sweep the arc i. Where this arc intersects the line f we term the point k, and from this point we lay off on the arc i 220 degrees, and draw the line $l\ e\ l'$, which we see coincides with the chord of the impulse face of the tooth. We set our dividers to the same radius by which we sweep the arc i and set one leg at the point b for a center and sweep the arc j'. If we measure this arc from the point j' to intersection of said arc j' with the line l we will find it to be sixty-four degrees, which accounts for our taking this number of degrees when we defined the face of our escape-wheel tooth, Fig. 129.

There is no reason why we should take twenty-degrees for the angle $k\ e\ l$ except that the practical construction of the larger sizes of cylinder watches has established the fact that this is about the right angle to employ, while in smaller watches it frequently runs up as high as twenty-five. Although the cylinder is seemingly a very simple escapement, it is really a very abstruce one to follow out so as to become familiar with all of its actions.

THE CYLINDER PROPER CONSIDERED.

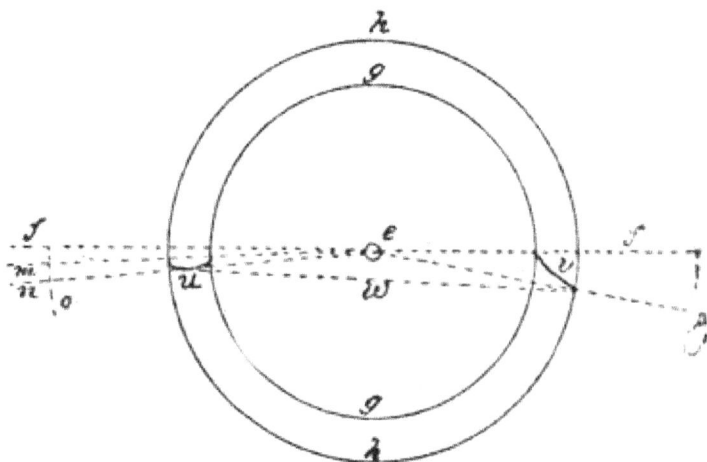

Fig. 131

We will now proceed and consider the cylinder proper, and to aid us in understanding the position and relation of the parts we refer to Fig. 131, where we repeat the circles d and h, shown in Fig. 130, which represents the inside and outside of the cylinder. We have here also repeated the line f of Fig. 130 as it cuts the cylinder in half, that is, divides it into two segments of 180 degrees each. If we conceive of a cylinder in which just one-half is cut away, that is, the lips are bounded by straight radial lines, we can also conceive of the relation and position of the parts shown in Fig. 130. The first position of which we should take cognizance is, the tooth D is moved back to the left so as to rest on the outside of our cylinder. The cylinder is also supposed to stand so that the lips correspond to the line f. On pressing the tooth D forward the incline of the tooth would attack the entrance lip of the cylinder at just about the center of the curved impulse face, imparting to the cylinder twenty degrees of angular motion, but the point of the tooth at d would exactly encounter the inner angle of the exit lip, and of course the cylinder would afford no rest for the tooth; hence, we see the importance of not cutting away too much of the half shell of the cylinder.

But before we further consider the action of the tooth *D* in its action as it passes the exit lip of the cylinder we must finish with the action of the tooth on the entrance lip. A very little thought and study of Fig. 130 will convince us that the incline of the tooth as it enters the cylinder will commence at *t*, Fig. 130, but at the close of the action the tooth parts from the lip on the inner angle. Now it is evident that it would require greater force to propel the cylinder by its inner angle than by the outer one. To compensate for this we round the edge of the entrance lip so that the action of the tooth instead of commencing on the outer angle commences on the center of the edge of the entrance lip and also ends its action on the center of the entrance lip. To give angular extent enough to the shell of the cylinder to allow for rounding and also to afford a secure rest for the tooth inside the cylinder, we add six degrees to the angular extent of the entrance lip of the cylinder shell, as indicated on the arc *o'*, Fig. 131, three of these degrees being absorbed for rounding and three to insure a dead rest for the tooth when it enters the cylinder.

WHY THE ANGULAR EXTENT IS INCREASED.

Without rounding the exit lip the action of the tooth on its exit would be entirely on the inner angle of the shell. To obviate this it is the usual practice to increase the angular extent of the cylinder ten degrees, as shown on the arc *o'* between the lines *f* and *p*, Fig. 131. Why we should allow ten degrees on the exit lip and but six degrees on the entrance lip will be understood by observing Fig. 130, where the radial lines *s* and *r* show the extent of angular motion of the cylinder, which would be lost if the tooth commenced to act on the inner angle and ended on the outer angle of the exit lip. This arc is a little over six degrees, and if we add a trifle over three degrees for rounding we would account for the ten degrees between the lines *f* and *p*, Fig. 131. It will now be seen that the angular extent is 196 degrees. If we draw the line *w* we can see in what proportion the measurement should be made between the outer diameter of the cylinder and the measure of the half shell. It will be seen on measurement that the distance between the center *e* and the line *w* is about one-fifteenth part of the outer diameter of the cylinder and consequently with a cylinder which measures 45/1000 of an inch in diameter, now the half shell should measure half of the entire diameter of the cylinder plus one-fifteenth part of such diameter, or 25-1/2 thousandths of an inch.

After these proportions are understood and the drawing made, the eye will get accustomed to judging pretty near what is required; but much the safer plan is to measure, where we have the proper tools for doing so. Most workmen have an idea that the depth or distance at which the cylinder is set from the escape wheel is a matter of adjustment; while this is true to a certain extent, still there is really only one position for the center of the cylinder, and that is so that the center of the pivot hole coincides exactly with the center of the chord to the curve of the impulse face of the tooth or the point e, Fig. 130. Any adjustment or moving back and forth of the chariot to change the depth could only be demanded where there was some fault existing in the cylinder or where it had been moved out of its proper place by some genius as an experiment in cylinder depths. It will be evident on observing the drawing at Fig. 131 that when the cylinder is performing an arc of vibration, as soon as the entrance lip has passed the point indicated by the radial line $e\ x$ the point of the escape-wheel tooth will commence to act on the cylinder lip and continue to do so through an arc of forty degrees, or from the lines x to l.

MAKING A WORKING MODEL.

To practically study the action of the cylinder escapement it is well to make a working model. It is not necessary that such a model should contain an entire escape wheel; all that is really required is two teeth cut out of brass of the proper forms and proportions and attached to the end of an arm 4-7/8" long with studs riveted to the U arms to support the teeth. This U arm is attached to the long arm we have just mentioned. A flat ring of heavy sheet brass is shaped to represent a short transverse section of a cylinder. This segment is mounted on a yoke which turns on pivots. In making such a model we can employ all the proportions and exact forms of the larger drawings made on a ten-inch radius. Such a model becomes of great service in learning the importance of properly shaping the lips of the cylinder. And right here we beg to call attention to the fact that in the ordinary repair shop the proper shape of cylinder lips is entirely neglected.

PROPER SHAPE OF CYLINDER LIPS.

The workman buys a cylinder and whether the proper amount is cut away from the half shell, or the lips, the correct form is entirely ignored, and still careful attention to the form of the cylinder lips adds full ten per cent. to the efficiency of the motive force as applied to the cylinder. In making study drawings of the cylinder escapement it is not necessary to employ paper so large that we can establish upon it the center of the arc which represents the periphery of our escape wheel, as we have at our disposal two plans by which this can be obviated. First, placing a bit of bristol board on our drawing-board in which we can set one leg of our dividers or compasses when we sweep the peripheral arc which we use in our delineations; second, making three arcs in brass or other sheet metal, viz.: the periphery of the escape wheel, the arc passing through the center of the chord of the arc of the impulse face of the tooth, and the arc passing through the point of the escape-wheel tooth. Of these plans we favor the one of sticking a bit of cardboard on the drawing board outside of the paper on which we are making our drawing.

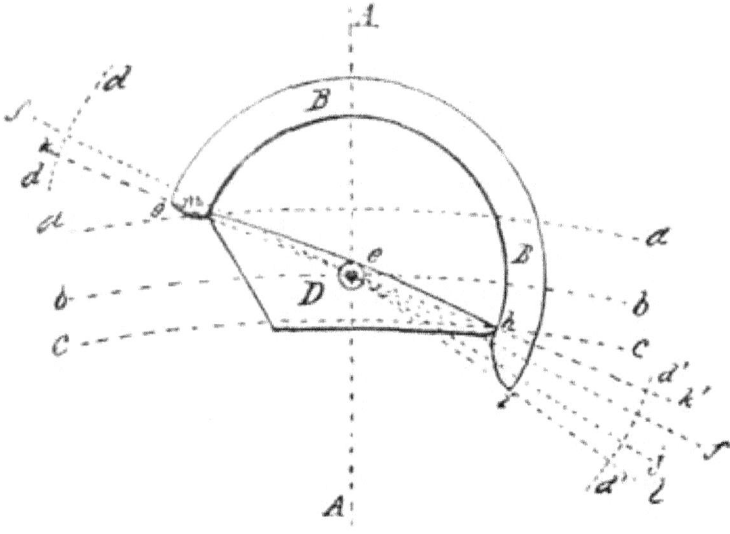

Fig. 132

At Fig. 132 we show the position and relation of the several parts just as the tooth passes into the shell of the cylinder, leaving the lip of the cylinder just as the tooth parted with it. The half shell of the cylinder as shown occupies 196 degrees or the larger arc embraced between the radial lines *k* and *l*. In drawing the entrance lip the acting face is made almost identical with a radial line except to round the corners for about one-third the thickness of the cylinder shell. No portion, however, of the lip can be considered as a straight line, but might be described as a flattened curve.

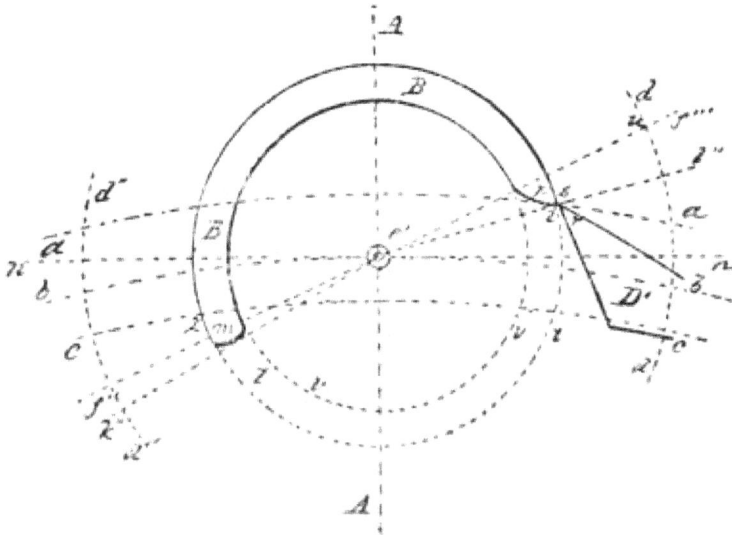

Fig. 133

A little study of what would be required to get the best results after making such a drawing will aid the pupil in arriving at the proper shape, especially when he remembers that the thickness of the cylinder shell of a twelve-line watch is only about five one-thousandths of an inch. But because the parts are small we should not shirk the problem of getting the most we possibly can out of a cylinder watch.

The extent of arc between the radial lines *k f*, as shown in Fig. 132, is four degrees. Although in former drawings we showed the angular extent added as six degrees, as we show the lip *m* in Fig. 132, two degrees are lost in rounding. The space *k f* on the egress or exit side is intended to be about four degrees, which shows the extent of lock. We show at Fig. 133 the tooth *D* just having passed out of the cylinder, having parted with the exit lip *p*.

In making this drawing we proceed as with Fig. 132 by establishing a center for our radius of 10" outside of our drawing paper and drawing the line *A A* to such center and sweeping the arcs *a b c*. We establish the point *e*, which represents the center of our cylinder, as before. We take the space to represent the radial extent of the outside of our cylinder in our dividers and from *e* as a center sweep a fine pencil line, represented by the dotted line *t* in our drawing; and where this circle intersects the arc *a* we name it the point *s*; and it is at this point the heel of our escape-wheel tooth must part with the exit lip of the cylinder. From *e* as a center and through the point *s* we draw the line *e l″*. With our dividers set to the radius of any convenient arc which we have divided into degrees, we sweep the short arc *d′*. The intersection of this arc with the line *e l″* we name the point *u*; and from *e* as a center we draw the radial line *e u f′*. We place the letter *f″* in connection with this line because it (the line) bears the same relations to the half shell of the cylinder shown in Fig. 133 that the line *f* does to the half shell (*D*) shown in Fig. 132. We draw the line *f″ f‴*, Fig. 133, which divides the cylinder into two segments of 180 degrees each. We take the same space in our dividers with which we swept the interior of the cylinder in Fig. 132 and sweep the circle *v*, Fig. 133. From *e* as a center we sweep the short arc *d″*, Fig. 133, and from its intersection of the line *f″* we lay off six degrees on said arc *d″* and draw the line *e′ k″*, which defines the angular extent of our entrance lip to the half shell of the cylinder in Fig. 133. We draw the full lines of the cylinder as shown.

We next delineate the heel of the tooth which has just passed out of the cylinder, as shown at *D′*, Fig. 133. We now have a drawing showing the position of the half shell of the cylinder just as the tooth has passed the exit lip. This drawing also represents the position of the half shell of the cylinder when the tooth rests against it on the outside. If we should make a drawing of an escape-wheel tooth shaped exactly as the one shown at Fig. 132 and the point of the tooth resting at *x*, we would show the position of a tooth encountering the cylinder after a tooth which has been engaged in the inside of the shell has passed out. By following the instructions now given, we can delineate a tooth in any of its relations with the cylinder shell.

DELINEATING AN ESCAPE-WHEEL TOOTH WHILE IN ACTION.

We will now go through the operation of delineating an escape-wheel tooth while in action. The position we shall assume is the one in which the cylinder and escape-wheel tooth are in the relation of the passage of half the impulse face of the tooth into the cylinder. To do this is simple enough: We first produce the arcs *a b c*, Fig. 133, as directed, and then proceed to delineate a tooth as in previous instances. To delineate our cylinder in the position we have assumed above, we take the space between the points *e d* in our dividers and setting one leg at *d* establish the point *g*, to represent the center of our cylinder. If we then sweep the circle *h* from the center of *g* we define the inner surface of the shell of our cylinder.

Strictly speaking, we have not assumed the position we stated, that is, the impulse face of the tooth as passing half way into the cylinder. To comply strictly with our statement, we divide the chord of the impulse face of the tooth *A* into eight equal spaces, as shown. Now as each of these spaces represent the thickness of the cylinder, if we take in our dividers four of these spaces and half of another, we have the radius of a circle passing the center of the cylinder shell. Consequently, if with this space in our dividers we set the leg at *d*, we establish on the arc *b* the point *i*. We locate the center of our cylinder when one-half of an entering tooth has passed into the cylinder. If now from the new center with our dividers set at four of the spaces into which we have divided the line *e f* we can sweep a circle representing the inner surface of the cylinder shell, and by setting our dividers to five of these spaces we can, from *i* as a center, sweep an arc representing the outside of the cylinder shell. For all purposes of practical study the delineation we show at Fig. 133 is to be preferred, because, if we carry out all the details we have described, the lines would become confused. We set our dividers at five of the spaces on the line *e f* and from *g* as a center sweep the circle *j*, which delineates the outer surface of our cylinder shell.

Let us now, as we directed in our former instructions, draw a flattened curve to represent the acting surface of the entrance lip of our cylinder as if it were in direct contact with the impulse face of the tooth. To delineate the exit lip we draw from the center *g*, Fig. 134, to the radial line *g k*, said line passing through the point of contact between the tooth and entrance lip of the cylinder. Let us next continue this line on the opposite side of the point *g*, as shown at *g k'*, and we thus bisect the cylinder shell into two equal parts of 180 degrees each.

151

As we previously explained, the entire extent of the cylinder half shell is 196 degrees. We now set our dividers to the radius of any convenient arc which we have divided into degrees, and from g as a center sweep the short arc l l, and from the intersection of this arc with the line g k' we lay off sixteen degrees on the said arc l and establish the point n, from g as a center draw the radial line g n'. Take ten degrees from the same parent arc and establish the point m, then draw the line g m'. Now the arc on the circles h j between the lines g n' and g m limits the extent of the exit lip of the cylinder and the arc between the lines g k' and g m' represents the locking surface of the cylinder shell.

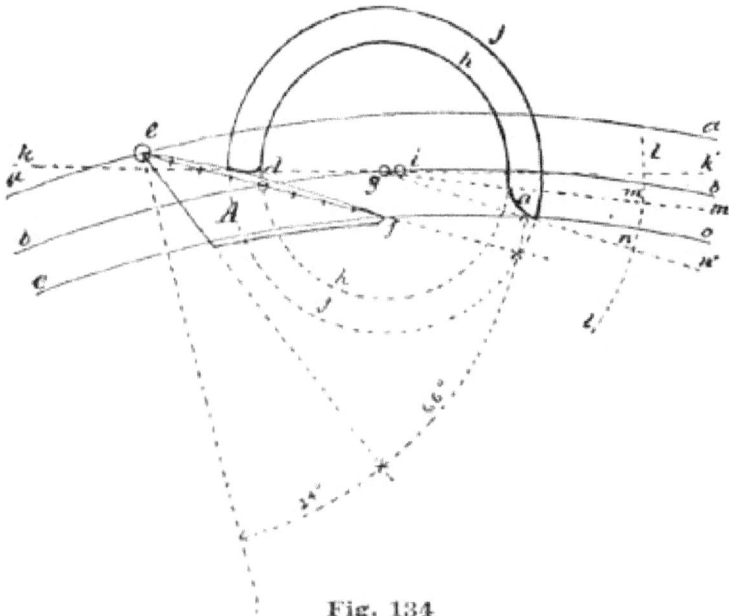

Fig. 134

To delineate the U arms we refer to Fig. 135. Here, again, we draw the arc a b c and delineate a tooth as before. From the point e located at the heel of the tooth we draw the radial line e e'. From the point e we lay off on the arc a five degrees and establish the point p; we halve this space and draw the short radial line p' s' and p s. From the point e on the arc A we lay off twenty-four degrees and establish the point t, which locates the heel of the next tooth in advance of A. At two and a half degrees to the right of the point t we locate the point r and draw the short radial line r s. On the arc b and half way between the lines p s and r s, we establish the point u, and from it as a center we sweep the arc v defining the curve of the U arms.

We have now given minute instructions for drawing a cylinder escapement in all its details except the extent of the banking slot of the cylinder, which is usually made to embrace an angular extent of 270 degrees; consequently, the pillar of the cylinder will not measure more than ninety degrees of angular extent.

There is no escapement constructed where carefully-made drawings tend more to perfect knowledge of the action than the cylinder. But it is necessary with the pupil to institute a careful analysis of the actions involved. In writing on a subject of this kind it is extremely perplexing to know when to stop; not that there is so much danger of saying too much as there is not having the words read with attention.

As an illustration, let us consider the subject of depth between the cylinder and the escape wheel. As previously stated, 196 degrees of cylinder shell should be employed; but suppose we find a watch in which the half shell has had too much cut away, so the tooth on entering the half shell after parting with the entrance lip does not strike dead on the inside of the shell, but encounters the edge of the exit lip. In this case the impulse of the balance would cause the tooth to slightly retrograde and the watch would go but would lack a good motion. In such an instance a very slight advance of the chariot would remedy the fault—not perfectly remedy it, but patch up, so to speak—and the watch would run.

In this day, fine cylinder watches are not made, and only the common kind are met with, and for this reason the student should familiarize himself with all the imaginary faults which could occur from bad construction. The best way to do this is to delineate what he (the student) knows to be a faulty escapement, as, for instance, a cylinder in which too much of the half shell is cut away; but in every instance let the tooth be of the correct form. Then delineate an escapement in which the cylinder is correct but the teeth faulty; also change the thickness of the cylinder shell, so as to make the teeth too short. This sort of practice makes the pupil think and study and he will acquire a knowledge which will never be forgotten, but always be present to aid him in the puzzles to which the practical watchmaker is every day subject.

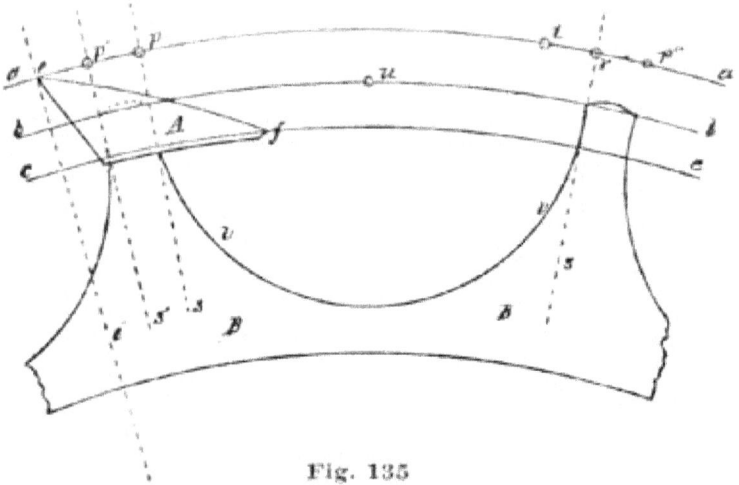

Fig. 135

The ability to solve these perplexing problems determines in a great degree the worth of a man to his employer, in addition to establishing his reputation as a skilled workman. The question is frequently asked, "How can I profitably employ myself in spare time?" It would seem that a watchmaker could do no better than to carefully study matters horological, striving constantly to attain a greater degree of perfection, for by so doing his earning capacity will undoubtedly be increased.

III. THE CHRONOMETER ESCAPEMENT

Undoubtedly "the detent," or, as it is usually termed, "the chronometer escapement," is the most perfect of any of our portable time measurers. Although the marine chronometer is in a sense a portable timepiece, still it is not, like a pocket watch, capable of being adjusted to positions. As we are all aware, the detent escapement is used in fine pocket watches, still the general feeling of manufacturers is not favorable to it. Much of this feeling no doubt is owing to the mechanical difficulties presented in repairing the chronometer escapements when the detent is broken, and the fact that the spring detent could not be adjusted to position. We shall have occasion to speak of position adjustments as relate to the chronometer escapement later on.

ADVANTAGES OF THE CHRONOMETER.

We will proceed now to consider briefly the advantages the detent escapement has over all others. It was soon discovered in constructing portable timepieces, that to obtain the best results the vibrations of the balance should be as free as possible from any control or influence except at such times as it received the necessary impulse to maintain the vibrations at a constant arc. This want undoubtedly led to the invention of the detent escapement. The early escapements were all frictional escapements, *i.e.*, the balance staff was never free from the influence of the train. The verge escapement, which was undoubtedly the first employed, was constantly in contact with the escape wheel, and was what is known as a "recoiling beat," that is, the contact of the pallets actually caused the escape wheel to recoil or turn back. Such escapements were too much influenced by the train, and any increase in power caused the timepiece to gain. The first attempt to correct this imperfection led to the invention and introduction of the fusee, which enabled the watchmaker to obtain from a coiled spring nearly equal power during the entire period of action. The next step in advance was the "dead-beat escapement," which included the cylinder and duplex. In these frictional escapements the balance staff locked the train while the balance performed its arc of vibration.

FRICTIONAL ESCAPEMENTS IN HIGH FAVOR.

These frictional escapements held favor with many eminent watchmakers even after the introduction of the detached escapements. It is no more than natural we should inquire, why? The idea with the advocates of the frictional rest escapements was, the friction of the tooth acted as a *corrective*, and led no doubt to the introduction of going-barrel watches. To illustrate, suppose in a cylinder watch we increase the motive power, such increase of power would not, as in the verge escapement, increase the rapidity of the vibrations; it might, in fact, cause the timepiece to run slower from the increased friction of the escape-wheel tooth on the cylinder; also, in the duplex escapement the friction of the locking tooth on the staff retards the vibrations.

Dr. Hooke, the inventor of the balance spring, soon discovered it could be manipulated to isochronism, *i.e.*, so arcs of different extent would be formed in equal time. Of course, the friction-rest escapement requiring a spring to possess different properties from one which would be isochronal with a perfectly detached escapement, these two frictional escapements also differing, the duplex requiring other properties from what would isochronize a spring for a cylinder escapement. Although pocket watches with duplex and cylinder escapements having balances compensated for heat and cold and balance springs adjusted to isochronism gave very good results, careful makers were satisfied that an escapement in which the balance was detached and free to act during the greater proportion of the arc of vibration and uncontrolled by any cause, would do still better, and this led to the detent escapement.

FAULTS IN THE DETENT ESCAPEMENT.

As stated previously, the detent escapement having pronounced faults in positions which held it back, it is probable it would never have been employed in pocket watches to any extent if it had not acquired such a high reputation in marine chronometers. Let us now analyze the influences which surround the detent escapement in a marine chronometer and take account of the causes which are combined to make it an accurate time measurer, and also take cognizance of other interfering causes which have a tendency to prevent desired results. First, we will imagine a balance with its spring such as we find in fine marine chronometers. It has small pivots running in highly-polished jewels; such pivots are perfectly cylindrical, and no larger than are absolutely necessary to endure the task imposed upon them—of carrying the weight of the balance and endure careful handling.

To afford the necessary vibrations a spring is fitted, usually of a helical form, so disposed as to cause the balance to vibrate in arcs back and forth in equal time, *provided these arcs are of equal extent*. It is now to be taken note of that we have it at our disposal and option to make these arcs equal in time duration, *i.e.*, to make the long or short arcs the quickest or to synchronize them. We can readily comprehend we have now established a very perfect measure of short intervals of time. We can also see if we provide the means of maintaining these vibrations and counting them we should possess the means of counting the flights of time with great accuracy. The conditions which surround our balance are very constant, the small pivots turning in fine hard jewels lubricated with an oil on which exposure to the action of the air has little effect, leaves but few influences which can interfere with the regular action of our balance. We add to the influences an adjustable correction for the disturbances of heat and cold, and we are convinced that but little could be added.

ANTAGONISTIC INFLUENCES.

In this combination we have pitted two antagonistic forces against each other, viz., the elasticity of the spring and the weight and inertia of the balance; both forces are theoretically constant and should produce constant results. The mechanical part of the problem is simply to afford these two forces perfect facilities to act on each other and compel each to realize its full effect. We must also devise mechanical means to record the duration of each conflict, that is, the time length of each vibration. Many years have been spent in experimenting to arrive at the best propositions to employ for the several parts to obtain the best practical results. Consequently, in designing a chronometer escapement we must not only draw the parts to a certain form, but consider the quality and weight of material to employ.

To illustrate what we have just said, suppose, in drawing an escape wheel, we must not only delineate the proper angle for the acting face of the tooth, but must also take cognizance of the thickness of the tooth. By thickness we mean the measurement of extent of the tooth in the direction of the axis of the escape wheel. An escape-wheel tooth might be of the best form to act in conveying power to the balance and yet by being too thin soon wear or produce excessive friction. How thick an escape wheel should be to produce best results, is one of the many matters settled only by actual workshop experience.

FACTORS THAT MUST BE CONSIDERED.

Even this experience is in every instance modified by other influences. To illustrate: Let us suppose in the ordinary to-day marine chronometer the escape-wheel teeth exerted a given average force, which we set down as so many grains. Now, if we should employ other material than hammer-hardened brass for an escape wheel it would modify the thickness; also, if we should decrease the motive power and increase the arc of impulse. Or, if we should diminish the extent of the impulse arc and add to the motive force, every change would have a controlling influence. In the designs we shall employ, it is our purpose to follow such proportions as have been adopted by our best makers, in all respects, including form, size and material. We would say, however, there has been but little deviation with our principal manufacturers of marine chronometers for the last twenty years as regards the general principle on which they were constructed, the chief aim being to excel in the perfection of the several parts and the care taken in the several adjustments.

Before we proceed to take up the details of constructing a chronometer escapement we had better master the names for the several parts. We show at Fig. 136 a complete plan of a chronometer escapement as if seen from the back, which is in reality the front or dial side of the "top plate." The chronometer escapement consists of four chief or principal parts, viz.: The escape wheel, a portion of which is shown at A; the impulse roller B; unlocking or discharging roller C, and the detent D. These principal parts are made up of sub-parts: thus, the escape wheel is composed of arms, teeth, recess and collet, the recess being the portion of the escape wheel sunk, to enable us to get wide teeth actions on the impulse pallet. The collet is a brass bush on which the wheel is set to afford better support to the escape wheel than could be obtained by the thinned wheel if driven directly on the pinion arbor. The impulse roller is composed of a cylindrical steel collet B, the impulse pallet d (some call it the impulse stone), the safety recess b b. The diameter of the impulse collet is usually one-half that of the escape wheel. This impulse roller is staked directly on the balance staff, and its perfection of position assured by resting against the foot of the shoulder to which the balance is secured. This will be understood by inspecting Fig. 137, which is a vertical longitudinal section of a chronometer balance staff, the lower side of the impulse roller being cupped out at c with a ball grinder and finished a ball polish.

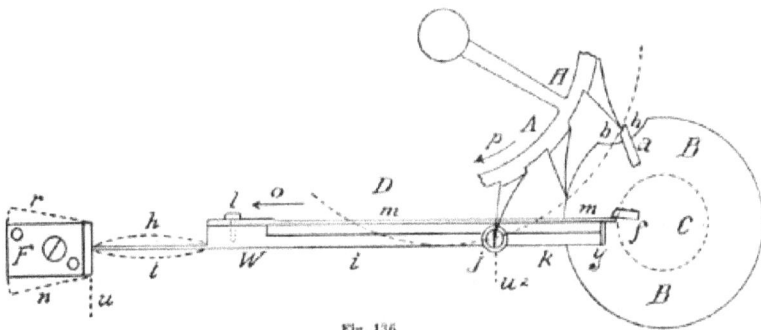

Fig. 136

It will be seen the impulse roller is staked flat against the hub E of the balance staff. The unlocking roller, or, as it is also called, the discharging roller, C, is usually thinner than the impulse roller and has a jewel similar to the impulse jewel a shown at f. This roller is fitted by friction to the lower part of the balance staff and for additional security has a pipe or short socket e which embraces the balance staff at g. The pipe e is usually flattened on opposite sides to admit of employing a special wrench for turning the discharging roller in adjusting the jewel for opening the escapement at the proper instant to permit the escape wheel to act on the impulse jewel *a*. The parts which go to make up the detent *D* consist of the "detent foot" *F*, the detent spring *h*, the detent blade *i*, the jewel pipe *j*, the locking jewel (or stone) *s*, the "horn" of the detent *k*, the "gold spring" (also called the auxiliary and lifting spring) *m*. This lifting or gold spring *m* should be made as light and thin as possible and stand careful handling.

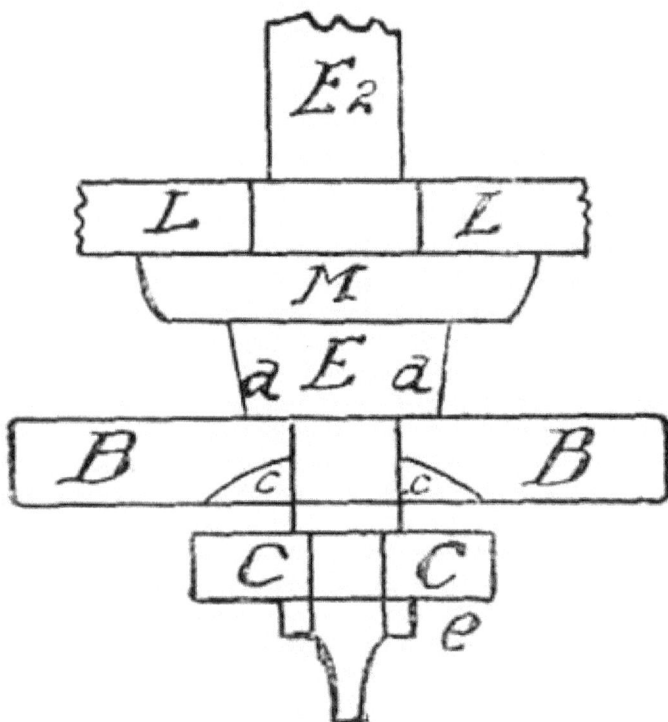

Fig. 137

We cannot impress on our readers too much the importance of making a chronometer detent light. Very few detents, even from the hands of our best makers, are as light as they might be. We should in such construction have very little care for clumsy workmen who may have to repair such mechanism. This feature should not enter into consideration.

We should only be influenced by the feeling that we are working for best results, and it is acting under this influence that we devote so much time to establishing a correct idea of the underlying principles involved in a marine chronometer, instead of proceeding directly to the drawing of such an escapement and give empirical rules for the length of this or the diameter of that. As, for instance, in finishing the detent spring *h*, suppose we read in text books the spring should be reduced in thickness, so that a weight of one pennyweight suspended from the pipe *j* will deflect the detent ¼". This is a rule well enough for people employed in a chronometer factory, but for the horological student such fixed rules (even if remembered) would be of small use. What the student requires is sound knowledge of the "whys," in order that he may be able to thoroughly master this escapement.

FUNCTIONS OF THE DETENT.

We can see, after a brief analysis of the principles involved, that the functions required of the detent *D* are to lock the escape wheel *A* and hold it while the balance performs its excursion, and that the detent or recovering spring *h* must have sufficient strength and power to perform two functions: (1) Return the locking stone *s* back to the proper position to arrest and hold the escape wheel; (2) the spring *h* must also be able to resist, without buckling or cockling, the thrust of the escape wheel, represented by the arrows *p o*. Now we can readily understand that the lighter we make the parts *i j k m*, the weaker the spring *h* can be. You say, perhaps, if we make it too weak it will be liable to buckle under the pressure of the escape wheel; this, in turn, will depend in a great measure on the condition of the spring *h*. Suppose we have it straight when we put it in position, it will then have no stress to keep it pressed to the holding, stop or banking screw, which regulates the lock of the tooth. To obtain this stress we set the foot *F* of the detent around to the position indicated by the dotted lines *r* and *n*, and we get the proper tension on the detent spring to effect the lock, or rather of the detent in time to lock the escape wheel; but the spring *h*, instead of being perfectly straight, is bent and consequently not in a condition to stand the thrust of the escape wheel, indicated by the arrows *o p*.

OBTAINING THE BEST CONDITIONS.

Now the true way to obtain the best conditions is to give the spring *h* a set curvature before we put it in place, and then when the detent is in the proper position the spring *h* will have tension enough on it to bring the jewel *s* against the stop screw, which regulates the lock, and still be perfectly straight. This matter is of so much importance that we will give further explanation. Suppose we bend the detent spring *h* so it is curved to the dotted line *t*, Fig. 136, and then the foot *F* would assume the position indicated at the dotted line *r*. We next imagine the foot *F* to be put in the position shown by the full lines, the spring *h* will become straight again and in perfect shape to resist the thrust of the escape wheel.

Little "ways and methods" like the above have long been known to the trade, but for some reason are never mentioned in our text books. A detent spring 2/1000" thick and 80/1000" wide will stand the thrust for any well-constructed marine chronometer in existence, and yet it will not require half a pennyweight to deflect it one-fourth of an inch. It is a good rule to make the length of the detent from the foot *F* to the center of the locking jewel pipe *j* equal to the diameter of the escape wheel, and the length of the detent spring *h* two-sevenths of this distance. The length of the horn *k* is determined by the graphic plan and can be taken from the plotted plan. The end, however, should approach as near to the discharging jewel as possible and not absolutely touch. The discharging (gold) spring *m* is attached to the blade *i* of the detent with a small screw *l* cut in a No. 18 hole of a Swiss plate. While there should be a slight increase in thickness in the detent blade at *w*, where the gold spring is attached, still it should be no more than to separate the gold spring *m* from the detent blade *i*.

IMPORTANT CONSIDERATIONS.

It is important the spring should be absolutely free and not touch the detent except at its point of attachment at *w* and to rest against the end of the horn *k*, and the extreme end of *k*, where the gold spring rests, should only be what we may term a dull or thick edge. The end of the horn *k* (shown at *y*) is best made, for convenience of elegant construction, square—that is, the part *y* turns at right angles to *k* and is made thicker than *k* and at the same time deeper; or, to make a

162

comparison to a clumsy article, y is like the head of a nail, which is all on one side. Some makers bend the horn k to a curve and allow the end of the horn to arrest or stop the gold spring; but as it is important the entire detent should be as light as possible, the square end best answers this purpose. The banking placed at j should arrest the detent as thrown back by the spring h at the "point of percussion." This point of percussion is a certain point in a moving mass where the greatest effort is produced and would be somewhere near the point x, in a bar G turning on a pivot at z, Fig. 138. It will be evident, on inspection of this figure, if the bar G was turning on the center z it would not give the hardest impact at the end v, as parts of its force would be expended at the center z.

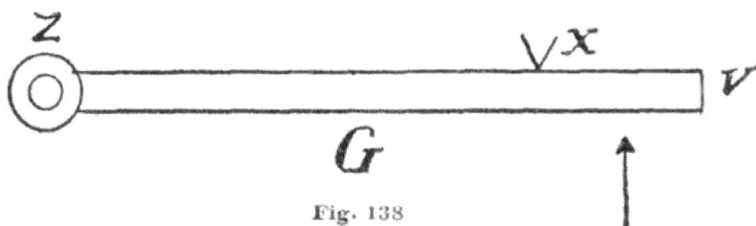

Fig. 138

DECISIONS ARRIVED AT BY EXPERIENCE.

Experience has decided that the impulse roller should be about half the diameter of the escape wheel, and experience has also decided that an escape wheel of fifteen teeth has the greatest number of advantages; also, that the balance should make 14,400 vibrations in one hour. We will accept these proportions and conditions as best, from the fact that they are now almost universally adopted by our best chronometer makers. Although it would seem as if these proportions should have established themselves earlier among practical men, we shall in these drawings confine ourselves to the graphic plan, considering it preferable. In the practical detail drawing we advise the employment of the scale given, *i.e.*, delineating an escape wheel 10" in diameter. The drawings which accompany the description are one-fourth of this size, for the sake of convenience in copying.

With an escape wheel of fifteen teeth the impulse arc is exactly twenty-four degrees, and of course the periphery of the impulse roller must intersect the periphery of the escape wheel for this arc

(24°). The circles *A B*, Fig. 139, represent the peripheries of these two mobiles, and the problem in hand is to locate and define the position of the two centers *a c*. These, of course, are not separated, the sum of the two radii, *i.e.*, 5" + 2-1/2" (in the large drawing), as these circles intersect, as shown at *d*. Arithmetically considered, the problem is quite difficult, but graphically, simple enough. After we have swept the circle *A* with a radius of 5", we draw the radial line *a f*, said line extending beyond the circle *A*.

LOCATING THE CENTER OF THE BALANCE STAFF.

Somewhere on this line is located the center of the balance staff, and it is the problem in hand to locate or establish this center. Now, it is known the circles which define the peripheries of the escape wheel and the impulse roller intersect at $e e^2$. We can establish on our circle *A* where these intersections take place by laying off twelve degrees, one-half of the impulse arc on each side of the line of centers *a f* on this circle and establishing the points $e e^2$. These points $e e^2$ being located at the intersection of the circles *A* and *B*, must be at the respective distances of 5" and 2-1/2" distance from the center of the circles *A B*; consequently, if we set our dividers at 2-1/2" and place one leg at *e* and sweep the short arc g^2, and repeat this process when one leg of the dividers is set at e^2, the intersection of the short arcs *g* and g^2 will locate the center of our balance staff. We have now our two centers established, whose peripheries are in the relation of 2 to 1.

To know, in the chronometer which we are supposed to be constructing, the exact distance apart at which to plant the hole jewels for our two mobiles, *i.e.*, escape wheel and balance staff, we measure carefully on our drawing the distance from *a* to *c* (the latter we having just established) and make our statement in the rule of three, as follows: As (10) the diameter of drawn escape wheel is to our real escape wheel so is the measured distance on our drawing to the real distance in the chronometer we are constructing.

It is well to use great care in the large drawing to obtain great accuracy, and make said large drawing on a sheet of metal. This course is justified by the degree of perfection to which measuring tools have arrived in this day. It will be found on measurement of the arc of the circle *B*, embraced between the intersections $e e^2$, that it is about forty-eight degrees. How much of this we can utilize in our escapement will depend very much on the perfection and accuracy of construction.

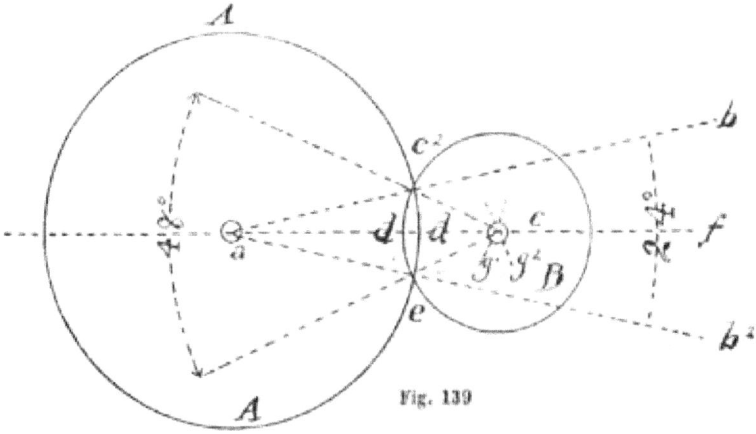

Fig. 139

We show at Fig. 140 three teeth of an escape wheel, together with the locking jewel E and impulse jewel D. Now, while theoretically we could commence the impulse as soon as the impulse jewel D was inside of the circle representing the periphery of the escape wheel, still, in practical construction, we must allow for contingencies. Before it is safe for the escape wheel to attack the impulse jewel, said jewel must be safely inside of said escape wheel periphery, in order that the attacking tooth shall act with certainty and its full effect. A good deal of thought and study can be bestowed to great advantage on the "action" of a chronometer escapement. Let us examine the conditions involved. We show in Fig. 140 the impulse jewel D just passing inside the circle of the periphery of the escape wheel. Now the attendant conditions are these: The escape wheel is locked fast and perfectly dead, and in the effort of unlocking it has to first turn backward against the effort of the mainspring; the power of force required for this effort is derived from the balance in which is stored up, so to speak, power from impulses imparted to the balance by former efforts of the escape wheel. In actual fact, the balance at the time the unlocking takes place is moving with nearly its greatest peripheral velocity and, as stated above, the escape wheel is at rest.

165

Here comes a very delicate problem as regards setting the unlocking or discharging jewel. Let us first suppose we set the discharging jewel so the locking jewel frees its tooth at the exact instant the impulse jewel is inside the periphery of the escape wheel. As just stated, the escape wheel is not only dead but actually moving back at the time the release takes place. Now, it is evident that the escape wheel requires an appreciable time to move forward and attack the impulse jewel, and during this appreciable time the impulse jewel has been moving forward inside of the arc *A A*, which represents the periphery of the escape wheel. The proper consideration of this problem is of more importance in chronometer making than we might at first thought have imagined, consequently, we shall dwell upon it at some length.

HOW TO SET THE DISCHARGING JEWEL.

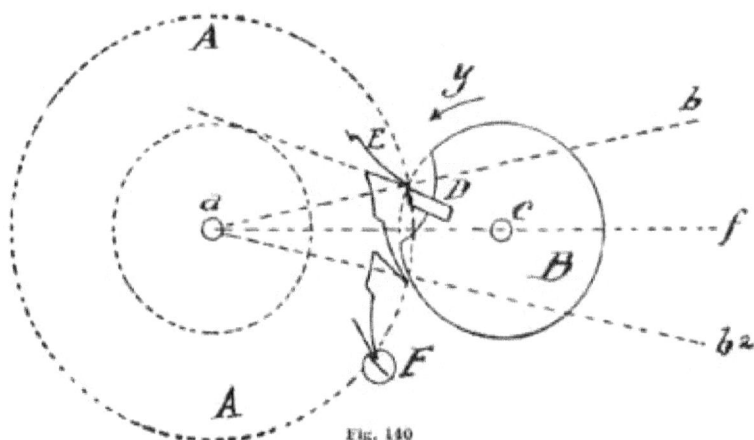

Fig. 140

Theoretically, the escape-wheel tooth should encounter the impulse jewel at the time—instant—both are moving with the same velocity. It is evident then that there can be no special rule given for this, *i.e.*, how to set the discharging jewel so it will free the tooth at exactly the proper instant, from the fact that one chronometer train may be much slower in getting to move forward from said train being heavy and clumsy in construction. Let us make an experiment with a real chronometer in illustration of our problem. To do so we remove our balance spring and place the balance in position. If we start the balance revolving in the direction of the arrow y, Fig. 140, it will cause the escapement to be unlocked and the balance to turn rapidly in one direction and with increasing velocity until, in fact, the escape wheel has but very little effect on the impulse jewel; in fact, we could, by applying some outside source of power—like blowing with a blow pipe on the balance—cause the impulse jewel to pass in advance of the escape wheel; that is, the escape-wheel tooth would not be able to catch the impulse jewel during the entire impulse arc. Let us suppose, now, we set our unlocking or discharging jewel in advance, that is, so the escapement is really unlocked a little before the setting parts are in the positions and relations shown in Fig. 141. Under the new conditions the escape wheel would commence to move and get sufficient velocity on it to act on the impulse jewel as soon as it was inside of the periphery of the escape wheel. If the balance was turned slowly now the tooth of the escape wheel would not encounter the impulse jewel at all, but fall into the passing hollow n; but if we give the balance a high velocity, the tooth would again encounter and act upon the jewel in the proper manner. Experienced adjusters of chronometers can tell by listening if the escape-wheel tooth attacks the impulse jewel properly, *i.e.*, when both are moving with similar velocities. The true sound indicating correct action is only given when the balance has its maximum arc of vibration, which should be about 1-1/4 revolutions, or perform an arc of 225 degrees on each excursion.

Fig. 142 is a side view of Fig. 141 seen in the direction of the arrow y. We have mentioned a chariot to which the detent is attached, but we shall make no attempt to show it in the accompanying drawings, as it really has no relation to the problem in hand; *i.e.*, explaining the action of the chronometer escapement, as the chariot relates entirely to the convenience of setting and adjusting the relation of the second parts. The size, or better, say, the inside diameter of the pipe at C, Fig. 143, which holds the locking jewel, should be about one-third of a tooth space, and the jewel made to fit perfectly. Usually, jewelmakers have a tendency to make this jewel too frail, cutting away the jewel back of the releasing angle (n, Fig. 143) too much.

A GOOD FORM OF LOCKING STONE.

A very practical form for a locking stone is shown in transverse section at Fig. 143. In construction it is a piece of ruby, or,

167

better, sapphire cut to coincide to its axis of crystallization, into first a solid cylinder nicely fitting the pipe *C* and finished with an after-grinding, cutting away four-tenths of the cylinder, as shown at *I*, Fig. 143. Here the line *m* represents the locking face of the jewel and the line *o* the clearance to free the escaping tooth, the angle at *n* being about fifty-four degrees. This angle (*n*) should leave the rounding of the stone intact, that is, the rounding of the angle should be left and not made after the flat faces *m o* are ground and polished. The circular space at *I* is filled with an aluminum pin. The sizes shown are of about the right relative proportions; but we feel it well to repeat the statement made previously, to the effect that the detent to a chronometer cannot well be made too light.

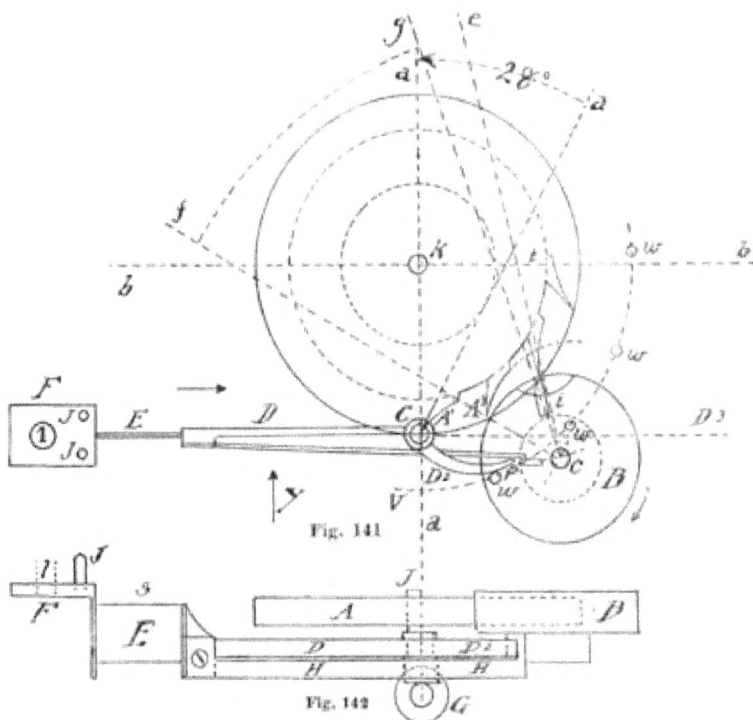

Fig. 141

Fig. 142

The so-called gold spring shown at *H*, Figs. 141 and 142, should also be as light as is consistent with due strength and can be made of the composite metal used for gold filled goods, as the only real benefit to be derived from employing gold is to avoid the necessity of applying oil to any part of the escapement. If such gold metal is employed, after hammering to obtain the greatest possible elasticity to the spring, the gold is filed away, except where the spring is acted upon by the discharging jewel *h*. We have previously mentioned the importance of avoiding wide, flat contacts between all acting surfaces, like where the gold spring rests on the horn of the detent at *p*; also where the detent banks on the banking screw, shown at *G*, Fig. 142. Under this principle the impact of the face of the discharging jewel with the end of the gold spring should be confined to as small a surface as is consistent with what will not produce abrasive action. The gold spring is shaped as shown at Fig. 142 and loses, in a measure, under the pipe of the locking jewel, a little more than one-half of the pipe below the blade of the detent being cut away, as shown in Fig. 143, where the lines *r r* show the extent of the part of the pipe which banks against the banking screw *G*. In this place even, only the curved surface of the outside of the pipe touches the screw *G*, again avoiding contact of broad surfaces.

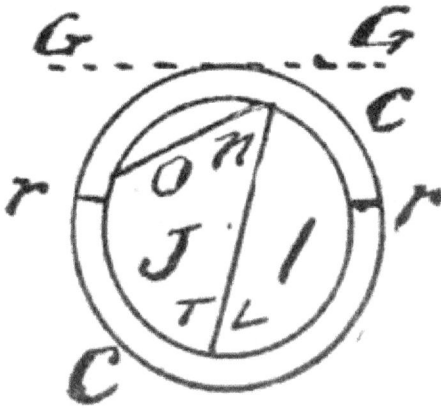

Fig. 143

We show the gold spring separate at Fig. 144. A slight torsion or twist is given to the gold spring to cause it to bend with a true curvature in the act of allowing the discharging pallet to pass back after unlocking. If the gold spring is filed and stoned to the right flexure, that is, the thinnest point properly placed or, say, located, the gold spring will not continue in contact with the discharging pallet any longer time or through a greater arc than during the process of unlocking. To make this statement better understood, let us suppose the weakest part of the gold spring H is opposite the arrow y, Fig. 141, it will readily be understood the contact of the discharging stone h would continue longer than if the point of greatest (or easiest) flexure was nearer to the pipe C. If the end D^2 of the horn of the detent is as near as it should be to the discharging stone there need be no fear but the escapement will be unlocked. The horn D^2 of the detent should be bent until five degrees of angular motion of the balance will unlock the escape, and the contact of discharging jewel h should be made without engaging friction. This condition can be determined by observing if the jewel seems to slide up (toward the pipe C) on the gold spring after contact. Some adjusters set the jewel J, Figs. 143 and 141, in such a way that the tooth rests close to the base; such adjusters claiming this course has a tendency to avoid cockling or buckling of the detent spring E. Such adjusters also set the impulse jewel slightly oblique, so as to lean on the opposite angle of the tooth. Our advice is to set both stones in places corresponding to the axis of the balance staff, and the escape-wheel mobiles.

THE DETENT SPRING.

Fig. 144

It will be noticed we have made the detent spring E pretty wide and extended it well above the blade of the detent. By shaping the detent in this way nearly all the tendency of the spring E to cockle is annulled. We would beg to add to what we said in regard to setting jewels obliquely. We are unable to understand the advantage of wide-faced stones and deep teeth when we do not take advantage of the wide surfaces which we assert are important. We guarantee that with a detent and spring made as we show, there will be no tendency to cockle, or if there is, it will be too feeble to even display itself. Those who have had extended experience with chronometers cannot fail to have noticed a gummy secretion which accumulates on the impulse and discharging stones of a chronometer, although no oil is ever applied to them. We imagine this coating is derived from the oil applied

170

to the pivots, which certainly evaporates, passes into vapor, or the remaining oil could not become gummy. We would advise, when setting jewels (we mean the locking, impulse and discharging jewels), to employ no more shellac than is absolutely necessary, depending chiefly on metallic contact for security.

DETAILS OF CONSTRUCTION.

We will now say a few words about the number of beats to the hour for a box or marine chronometer to make to give the best results. Experience shows that slow but most perfect construction has settled that 14,400, or four vibrations of the balance to a second, as the proper number, the weight of balance, including balance proper and movable weights, to be about 5-1/2 pennyweights, and the compensating curb about 1-2/10" in diameter. The escape wheel, 55/100" in diameter and recessed so as to be as light as possible, should have sufficient strength to perform its functions properly. The thickness or, more properly, the face extent of the tooth, measured in the direction of the axis of the escape wheel, should be about 1/20". The recessing should extend half way up the radial back of the tooth at t. The curvature of the back of the teeth is produced with the same radii as the impulse roller. To locate the center from which the arc which defines the back of the teeth is swept, we halve the space between the teeth A^2 and a^4 and establish the point n, Fig. 141, and with our dividers set to sweep the circle representing the impulse roller, we sweep an arc passing the point of the tooth A^3 and u, thus locating the center w. From the center k of the escape wheel we sweep a complete circle, a portion of which is represented by the arc $w\ v$. For delineating other teeth we set one leg of our dividers to agree with the point of the tooth and the other leg on the circle $w\ v$ and produce an arc like $z\ u$.

ORIGINAL DESIGNING OF THE ESCAPEMENT.

On delineating our chronometer escapement shown at Fig. 141 we have followed no text-book authority, but have drawn it according to such requirements as are essential to obtain the best results. An escapement of any kind is only a machine, and merely requires in its construction a combination of sound mechanical principles. Neither Saunier nor Britten, in their works, give instructions for drawing this escapement which will bear close analysis. It is not our intention, however, to criticise these authors, except we can present better methods and give correct systems.

TANGENTIAL LOCKINGS.

It has been a matter of great contention with makers of chronometer and also lever escapements as to the advantages of "tangential lockings." By this term is meant a locking the same as is shown at C, Fig. 141, and means a detent planted at right angles to a line radial to the escape-wheel axis, said radial line passing through the point of the escape-wheel tooth resting on the locking jewel. In escapements not set tangential, the detent is pushed forward in the direction of the arrow x about half a tooth space. Britten, in his "Hand-Book," gives a drawing of such an escapement. We claim the chief advantage of tangential locking to lie in the action of the escape-wheel teeth, both on the impulse stone and also on the locking stone of the detent. Saunier, in his "Modern Horology," gives the inclination of the front fan of the escape-wheel teeth as being at an angle of twenty-seven degrees to a radial line. Britten says twenty degrees, and also employs a non-tangential locking.

Our drawing is on an angle of twenty-eight degrees, which is as low as is safe, as we shall proceed to demonstrate. For establishing the angle of an escape-wheel tooth we draw the line $C\ d$, from the point of the escape-wheel tooth resting on the locking stone shown at C at an angle of twenty-eight degrees to radial line $C\ k$. We have already discussed how to locate and plant the center of the balance staff.

We shall not show in this drawing the angular motion of the escape wheel, but delineate at the radial lines $c\ e$ and $c\ f$ of the arc of the balance during the extent of its implication with the periphery of the escape wheel, which arc is one of about forty-eight degrees. Of this angle but forty-three degrees is attempted to be utilized for the purpose of impulse, five degrees being allowed for the impulse jewel to pass inside of the arc of periphery of the escape wheel before the locking jewel releases the tooth of the escape wheel resting upon it. At this point it is supposed the escape wheel attacks the impulse jewel, because, as we just explained, the locking jewel has released the tooth engaging it. Now, if the train had no weight, no inertia to overcome, the escape wheel tooth A^2 would move forward and attack the impulse

172

pallet instantly; but, in fact, as we have already explained, there will be an appreciable time elapse before the tooth overtakes the rapidly-moving impulse jewel. It will, of course, be understood that the reference letters used herein refer to the illustrations that have appeared on preceding pages.

If we reason carefully on the matter, we will readily comprehend that we can move the locking jewel, *i.e.*, set it so the unlocking will take place in reality before the impulse jewel has passed through the entire five degrees of arc embraced between the radial lines *c e* and *c g*, Fig. 141, and yet have the tooth attack the jewel after the five degrees of arc. In practice it is safe to set the discharging jewel *h* so the release of the held tooth A^1 will take place as soon as the tooth A^2 is inside the principal line of the escape wheel. As we previously explained, the contact between A^2 and the impulse jewel *i* would not in reality occur until the said jewel *i* had fully passed through the arc (five degrees) embraced between the radial lines *c e* and *c g*.

At this point we will explain why we drew the front fan of the escape-wheel teeth at the angle of twenty-eight degrees. If the fan of impulse jewel *i* is set radial to the axis of the balance, the engagement of the tooth A^2 would be at a disadvantage if it took place prior to this jewel passing through an arc of five degrees inside the periphery of the escape wheel. It will be evident on thought that if an escape-wheel tooth engaged the impulse stone before the five-degrees angle had passed, the contact would not be on its flat face, but the tooth would strike the impulse jewel on its outer angle. A continued inspection will also reveal the fact that in order to have the point of the tooth engage the flat surface of the impulse pallet the impulse jewel must coincide with the radial line *c g*. If we seek to remedy this condition by setting the impulse jewel so the face is not radial, but inclined backward, we encounter a bad engaging friction, because, during the first part of the impulse action, the tooth has to slide up the face of the impulse jewel. All things considered, the best action is obtained with the impulse jewel set so the acting face is radial to the balance staff and the engagement takes place between the tooth and the impulse jewel when both are moving with equal velocities, *i.e.*, when the balance is performing with an arc (or motion) of 1-1/4 revolutions or 225 degrees each way from a point of rest. Under such conditions the actual contact will not take place before some little time after the impulse jewel has passed the five-degree arc between the lines *c e* and *c g*.

THE DROP AND DRAW CONSIDERED.

Exactly how much drop must be allowed from the time the tooth leaves the impulse jewel before the locking tooth engages the locking jewel will depend in a great measure on the perfection of workmanship, but should in no instance be more than what is absolutely required to make the escapement safe. The amount of draw given to the locking stone c is usually about twelve degrees to the radial line $k\ a$. Much of the perfection of the chronometer escapement will always depend on the skill of the escapement adjuster and not on the mechanical perfection of the parts.

The jewels all have to be set by hand after they are made, and the distance to which the impulse jewel protrudes beyond the periphery of the impulse roller is entirely a matter for hand and eye, but should never exceed 2/1000". After the locking jewel c is set, we can set the foot F of the detent D forward or back, to perfect and correct the engagement of the escape-wheel teeth with the impulse roller B. If we set this too far forward, the tooth A^3 will encounter the roller while the tooth A^2 will be free.

We would beg to say here there is no escape wheel made which requires the same extreme accuracy as the chronometer, as the tooth spaces and the equal radial extent of each tooth should be only limited by our powers toward perfection. It is usual to give the detent a locking of about two degrees; that is, it requires about two degrees to open it, counting the center of fluxion of the detent spring E and five degrees of balance arc.

FITTING UP OF THE FOOT.

Several attempts have been made by chronometer makers to have the foot F adjustable; that is, so it could be moved back and forth with a screw, but we have never known of anything satisfactory being accomplished in this direction. About the best way of fitting up the foot F seems to be to provide it with two soft iron steady pins (shown at j) with corresponding holes in the chariot, said holes being conically enlarged so they (the pins) can be bent and manipulated so the detent not only stands in the proper position as regards the escape wheel, but also to give the detent spring E the proper elastic force to return in time to afford a secure locking to the arresting tooth of the escape wheel after an impulse has been given.

If these pins *j* are bent properly by the adjuster, whoever afterwards cleans the chronometer needs only to gently push the foot *F* forward so as to cause the pins *j* to take the correct positions as determined by the adjuster and set the screw *l* up to hold the foot *F* when all the other relations are as they should be, except such as we can control by the screw *G*, which prevents the locking jewel from entering too deeply into the escape wheel.

In addition to being a complete master of the technical part of his business, it is also desirable that the up-to-date workman should be familiar with the subject from a historical point of view. To aid in such an understanding of the matter we have translated from "L'Almanach de l'Horologerie et de la Bijouterie" the matter contained in the following chapter.

IV. HISTORY OF ESCAPEMENTS

.

It could not have been long after man first became cognizant of his reasoning faculties that he began to take more or less notice of the flight of time. The motion of the sun by day and of the moon and stars by night served to warn him of the recurring periods of light and darkness. By noting the position of these stellar bodies during his lonely vigils, he soon became proficient in roughly dividing up the cycle into sections, which he denominated the hours of the day and of the night. Primitive at first, his methods were simple, his needs few and his time abundant. Increase in numbers, multiplicity of duties, and division of occupation began to make it imperative that a more systematic following of these occupations should be instituted, and with this end in view he contrived, by means of burning lights or by restricting the flowing of water or the falling of weights, to subdivide into convenient intervals and in a tolerably satisfactory manner the periods of light.

These modest means then were the first steps toward the exact subdivisions of time which we now enjoy. Unrest, progress, discontent with things that be, we must acknowledge, have, from the appearance of the first clock to the present hour, been the powers which have driven on the inventive genius of watch and clockmakers to designate some new and more acceptable system for regulating the course of the movement. In consequence of this restless search after the best, a very considerable number of escapements have been invented and made up, both for clocks and watches; only a few, however, of the almost numberless systems have survived the test of time and been adopted in the manufacture of the timepiece as we know it now. Indeed, many such inventions never passed the experimental stage, and yet it would be very interesting to the professional horologist, the apprentice and even the layman to become more intimately acquainted with the vast variety of inventions made upon this domain since the inception of horological science. Undoubtedly, a complete collection of all the escapements invented would constitute a most instructive work for the progressive watchmaker, and while we are waiting for a competent author to take such an exhaustive work upon his hands, we shall endeavor to open the way and trust that a number of voluntary collaborators will come forward and assist us to the extent of their ability in filling up the chinks.

PROBLEMS TO BE SOLVED.

The problem to be solved by means of the escapement has always been to govern, within limits precise and perfectly regular, if it be possible, the flow of the motive force; that means the procession of the wheel-work and, as a consequence, of the hands thereto attached. At first blush it seems as if a continually-moving governor, such as is in use on steam engines, for example, ought to fulfil the conditions, and attempts have accordingly been made upon this line with results which have proven entirely unsatisfactory.

Having thoroughly sifted the many varieties at hand, it has been finally determined that the only means known to provide the most regular flow of power consists in intermittently interrupting the procession of the wheel-work, and thereby gaining a periodically uniform movement. Whatever may be the system or kind of escapement employed, the functioning of the mechanism is characterized by the suspension, at regular intervals, of the rotation of the last wheel of the train and in transmitting to a regulator, be it a balance or a pendulum, the power sent into that wheel.

ESCAPEMENT THE MOST ESSENTIAL PART.

Of all the parts of the timepiece the escapement is then the most essential; it is the part which assures regularity in the running of the watch or clock, and that part of parts that endows the piece with real value. The most perfect escapement would be that one which should perform its duty with the least influence upon the time of oscillation or vibration of the regulating organ. The stoppage of the train by the escapement is brought about in different ways, which may be gathered under three heads or categories. In the two which we shall mention first, the stop is effected directly upon the axis of the regulator, or against a piece which forms a part of that axis; the tooth of the escape wheel at the moment of its disengagement remains supported upon or against that stop.

In the first escapement invented and, indeed, in some actually employed to-day for certain kinds of timekeepers, we notice during the locking a retrograde movement of the escape wheel; to this kind of movement has been given the name of *recoil escapement*. It was recognized by the fraternity that this recoil was prejudicial to the regularity of the running of the mechanism and, after the invention of the pendulum and the spiral, inventive makers succeeded in replacing this sort of escapement with one which we now call the *dead-beat escapement*. In this latter the wheel, stopped by the axis of the regulator, remains immovable up to the instant of its disengagement or unlocking.

In the third category have been collected all those forms of escapement wherein the escape wheel is locked by an intermediate piece, independent of the regulating organ. This latter performs its vibrations of oscillation quite without interference, and it is only in contact with the train during the very brief moment of impulse which is needful to keep the regulating organ in motion. This category constitutes what is known as the *detached escapement* class.

Of the *recoil escapement* the principal types are: the *verge escapement* or *crown-wheel escapement* for both watches and clocks, and the *recoil anchor escapement* for clocks. The *cylinder* and *duplex escapements* for watches and the *Graham anchor escapement* for clocks are styles of the *dead-beat escapement* most often employed. Among the *detached escapements* we have the *lever* and *detent* or *chronometer escapements* for watches; for clocks there is no fixed type of detached lever and it finds no application to-day.

THE VERGE ESCAPEMENT.

The *verge escapement*, called also the *crown-wheel escapement*, is by far the simplest and presents the least difficulty in construction. We regret that the world does not know either the name of its originator nor the date at which the invention made its first appearance, but it seems to have followed very closely upon the birth of mechanical horology.

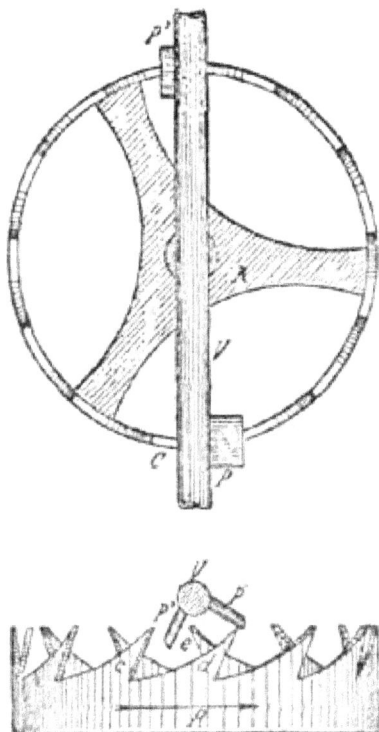

Figs. 145 and 146

Up to 1750 it was employed to the exclusion of almost all the others. In 1850 a very large part of the ordinary commercial watches were still fitted with the verge escapement, and it is still used under the form of *recoil anchor* in clocks, eighty years after the invention of the cylinder escapement, or in 1802. Ferdinand Berthoud, in his "History of the Measurement of Time," says of the balance-wheel escapement: "Since the epoch of its invention an infinite variety of escapements have been constructed, but the one which is employed in ordinary watches for every-day use is still the best." In referring to our illustrations, we beg first to call attention to the plates marked Figs. 145 and 146. This plate gives us two views of a verge escapement; that is, a

179

balance wheel and a verge formed by its two opposite pallets. The views are intentionally presented in this manner to show that the verge V may be disposed either horizontally, as in Fig. 146, or vertically, as in Fig. 145.

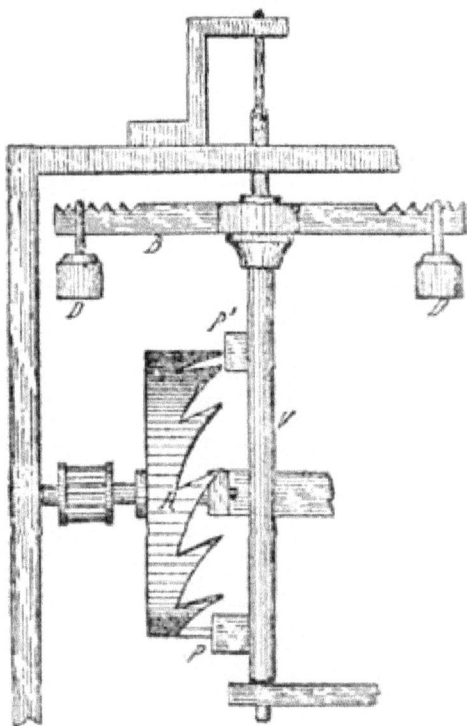

Fig. 147

Let us imagine that our drawing is in motion, then will the tooth d, of the crown wheel R, be pushing against the pallet P, and just upon the point of slipping by or escaping, while the opposite tooth e is just about to impinge upon the advancing pallet P'. This it does, and will at first, through the impulse received from the tooth d be forced back by the momentum of the pallet, that is, suffer a recoil; but on the return journey of the pallet P', the tooth e will then add its impulse to the receding pallet. The tooth e having thus accomplished its mission, will now slip by and the tooth c will come in lock with the pallet P

and, after the manner just described for *e*, continue the escapement. Usually these escape wheels are provided with teeth to the number of 11, 13 or 15, and always uneven. A great advantage possessed by this form of escapement is that it does not require any oil, and it may be made to work even under very inferior construction.

OLDEST ARRANGEMENT OF A CROWN-WHEEL ESCAPEMENT.

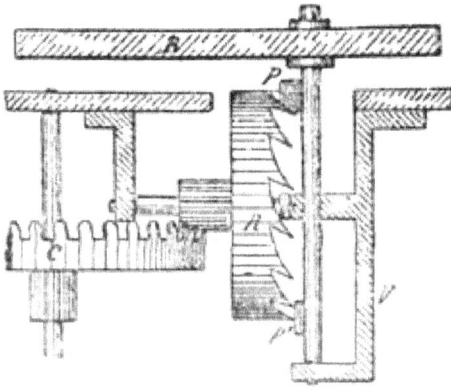

Fig. 148

Plate 147 shows us the oldest known arrangement of a crown-wheel escapement in a clock. *R* is the crown wheel or balance wheel acting upon the pallets *P* and *P'*, which form part of the verge *V*. This verge is suspended as lightly as possible upon a pliable cord *C* and carries at its upper end two arms, *B* and *B*, called adjusters, forming the balance. Two small weights *D D*, adapted to movement along the rules or adjusters serve to regulate the duration of a vibration. In Fig. 148 we have the arrangement adopted in small timepieces and watches: *B* represents the regulator in the form of a circular balance, but not yet furnished with a spiral regulating spring; *c* is the last wheel of the train and called the *fourth wheel*, it being that number distant from the great wheel. As will be seen, the verge provided with its pallets is vertically placed, as in the preceding plate.

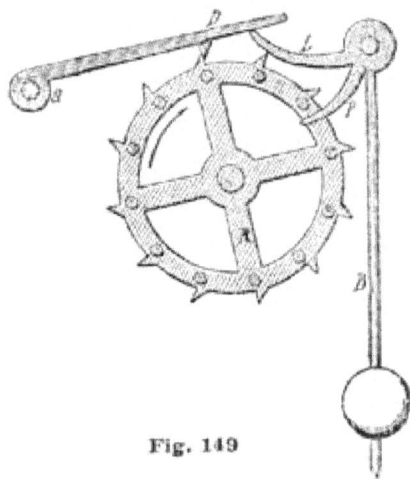

Fig. 149

Here it will quickly be seen that regarded from the standpoint of regularity of motion, this arrangement can be productive of but meager results. Subjected as it is to the influence of the slightest variation in the motive power and of the least jar or shaking, a balance wheel escapement improvided with a regulator containing within itself a regulating force, could not possibly give forth anything else than an unsteady movement. However, mechanical clocks fitted with this escapement offer indisputable advantages over the ancient clepsydra; in spite of their imperfections they rendered important services, especially after the striking movement had been added. For more than three centuries both this crude escapement and the cruder regulator were suffered to continue in this state without a thought of improvement; even in 1600, when Galileo discovered the law governing the oscillation of the pendulum, they did not suspect how important this discovery was for the science of time measurement.

GALILEO'S EXPERIMENTS.

Galileo, himself, in spite of his genius for investigation, was so engrossed in his researches that he could not seem to disengage the simple pendulum from the compound pendulums to which he devoted his attention; besides, he attributed to the oscillation an absolute generality of isochronism, which they did not possess; nor did he know how to apply his famous discovery to the measurement of time. In fact, it was not till after more than half a century had elapsed, in 1657, to be exact, that the celebrated Dutch mathematician and astronomer, Huygens, published his memoirs in which he made known to the world the degree of perfection which would accrue to clocks if the pendulum were adopted to regulate their movement.

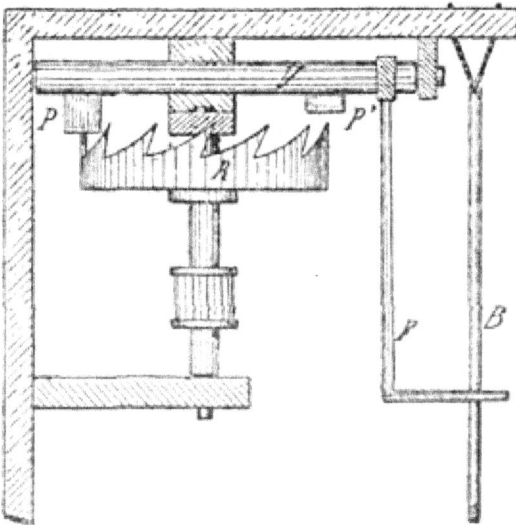

Fig. 150

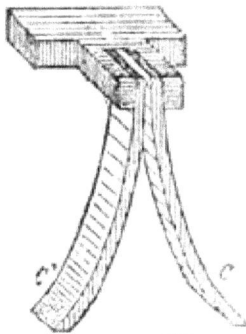

Fig. 151

An attempt was indeed made to snatch from Huygens and confer upon Galileo the glory of having first applied the pendulum to a clock, but this attempt not having been made until some time after the publication of "Huygens' Memoirs," it was impossible to place any faith in the contention. If Galileo had indeed solved the beautiful problem, both in the conception and the fact, the honor of the discovery was lost to him by the laziness and negligence of his pupil, Viviani, upon whom he had placed such high hopes. One thing is certain, that the right of priority of the discovery and the recognition of the entire world has been incontestably bestowed upon Huygens. The escapement which Galileo is supposed to have conceived and to which he applied the pendulum, is shown in Fig. 149. The wheel R is supplied with teeth, which lock against the piece D attached to a lever pivoted at a, and also with pins calculated to impart impulses to the pendulum through the pallet P. The arm L serves to disengage or unlock the wheel by lifting the lever D upon the return oscillation of the pendulum.

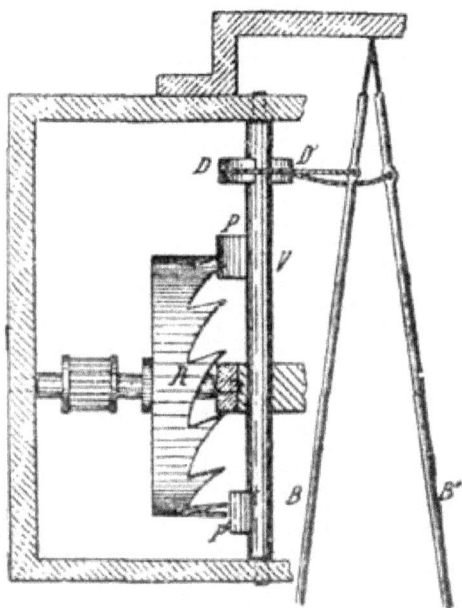

Fig. 152

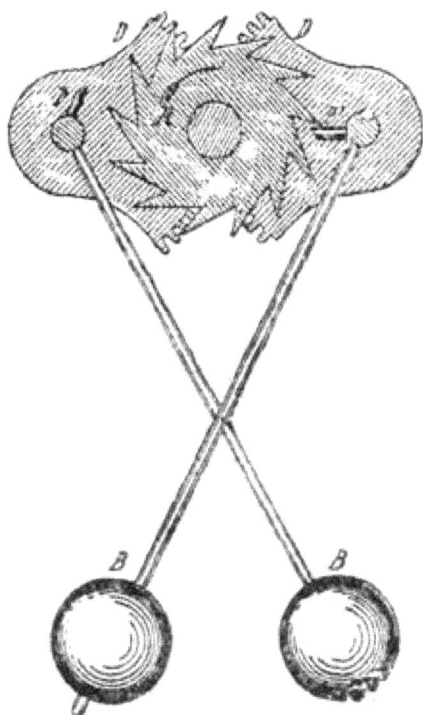

Fig. 153

A careful study of Fig. 150 will discover a simple transposition which it became necessary to make in the clocks, for the effectual adaptation of the pendulum to their regulation. The verge *V* was set up horizontally and the pendulum *B*, suspended freely from a flexible cord, received the impulses through the intermediation of the forked arm *F*, which formed a part of the verge. At first this forked arm was not thought of, for the pendulum itself formed a part of the verge. A far-reaching step had been taken, but it soon became apparent that perfection was still a long way off. The crown-wheel escapement forcibly incited the pendulum to wider oscillations; these oscillations not being as Galileo had believed, of unvaried durations, but they varied sensibly with the intensity of the motive power.

185

THE ATTAINMENT OF ISOCHRONISM BY HUYGENS.

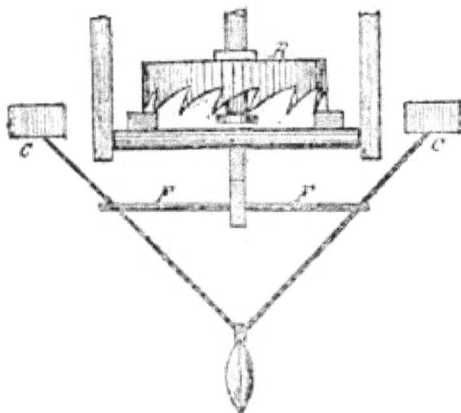

Fig. 154

Huygens rendered his pendulum *isochronous*; that is, compelled it to make its oscillations of equal duration, whatever might be the arc described, by suspending the pendulum between two metallic curves *c c'*, each one formed by an arc of a cycloid and against which the suspending cord must lie upon each forward or backward oscillation. We show this device in Fig. 151. In great oscillations, and by that we mean oscillations under a greater impulse, the pendulum would thus be shortened and the shortening would correct the time of the oscillation. However, the application of an exact cycloidal arc was a matter of no little difficulty, if not an impossibility in practice, and practical men began to grope about in search of an escapement which would permit the use of shorter arcs of oscillation. At London the horologist, G. Clement, solved the problem in 1675 with his rack escapement and recoil anchor. In the interval other means were invented, especially the addition of a second pendulum to correct the irregularities of the first. Such an escapement is pictured in Fig. 152. The verge is again vertical and carries near its upper end two arms *D D*, which are each connected by a cord with a pendulum. The two pendulums oscillate constantly in the inverse sense the one to the other.

ANOTHER TWO-PENDULUM ESCAPEMENT.

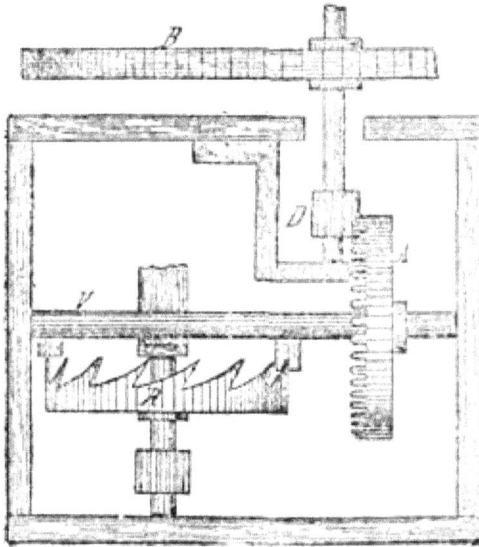

Fig. 155

We show another escapement with two pendulums in Fig. 153. These are fixed directly upon two axes, each one carrying a pallet P P' and a segment of a toothed wheel D D, which produces the effect of solidarity between them. The two pendulums oscillate inversely one to the other, and one after the other receives an impulse. This escapement was constructed by Jean Baptiste Dutertre, of Paris.

Fig. 154 shows another disposition of a double pendulum. While the pendulum here is double, it has but one bob; it receives the impulse by means of a double fork F. C C represents the cycloidal curves and are placed with a view of correcting the inequality in the duration of the oscillations. In watches the circular balances did not afford any better results than the regulating rods or rules of the clocks, and the pendulum, of course, was out of the question altogether; it therefore became imperative to invent some other regulating system.

187

Fig. 156

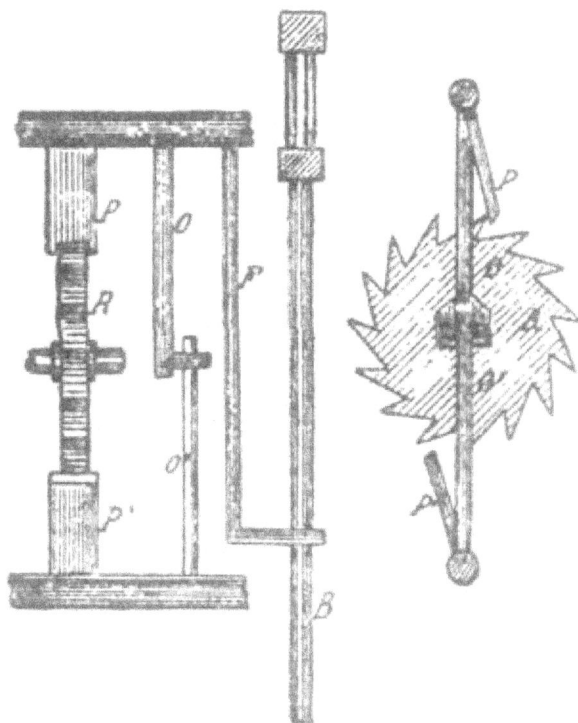

Fig. 157

It occured to the Abbé d'Hautefeuille to form a sort of resilient mechanism by attaching one end of a hog's bristle to the plate and the other to the balance near the axis. Though imperfect in results, this was nevertheless a brilliant idea, and it was but a short step to replace the bristle with a straight and very flexible spring, which later was supplanted by one coiled up like a serpent; but in spite of this advancement, the watches did not keep much better time. Harrison, the celebrated English horologist, had recourse to two artifices, of which the one consisted in giving to the pallets of the escapement such a curvature that the balance could be led back with a velocity corresponding to the extension of the oscillation; the second consisted of an accessory piece, the resultant action of which was analogous to that of the cycloidal curves in connection with the pendulum.

CORRECTING IRREGULARITIES IN THE VERGE ESCAPEMENT.

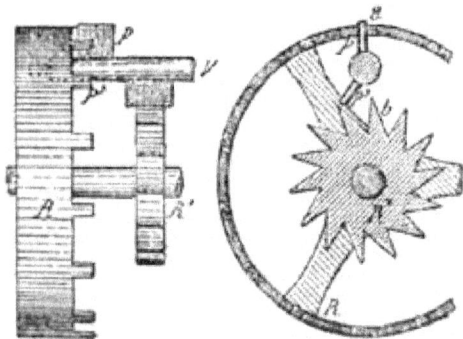

Fig. 158

Huygens attempted to correct these irregularities in the verge escapement in watches by amplifying the arc of oscillation of the balance itself. He constructed for that purpose a pirouette escapement shown in Fig. 155, in which a toothed wheel A adjusted upon the verge V serves as an intermediary between that and the balance B, upon the axis of which was fixed a pinion D. By this method he obtained extended arcs of vibration, but the vibrations were, as a consequence, very slow, and they still remained subject to all the irregularities arising from the variation in the motive power as well as from shocks. A little later, but about the same epoch, a certain Dr. Hook, of the Royal Society of London, contrived another arrangement by means of which he succeeded, so it appeared to him at least, in greatly diminishing the influence of shock upon the escapement; but many other, perhaps greater, inconveniences caused his invention to be speedily rejected. We shall give our readers an idea of what Dr. Hook's escapement was like.

On looking at Fig. 156 we see the escape wheel R, which was flat and in the form of a ratchet; it was provided with two balances. B B engaging each other in teeth, each one carrying a pallet P P' upon its axis; the axes of the three wheels being parallel. Now, in our drawing, the tooth a of the escape wheel exerts its lift upon the pallet P'; when this tooth escapes the tooth b will fall upon the pallet P' on the opposite side, a recoil will be produced upon the action of the two united balances, then the tooth b will give its impulse in the contrary direc-

190

tion. Considerable analogy exists between this form of escapement and that shown in Fig. 153 and intended for clocks. This was the busy era in the watchmaker's line. All the great heads were pondering upon the subject and everyone was on the *qui vive* for the newest thing in the art.

Fig. 159

In 1674 Huygens brought out the first watch having a regulating spring in the form of a spiral; the merit of this invention was disputed by the English savant, Dr. Hook, who pretended, as did Galileo, in the application of the pendulum, to have priority in the idea. Huygens, who had discovered and corrected the irregularities in the oscillations of the pendulum, did not think of those of the balance with the spiral spring. And it was not until the close of the year 1750 that Pierre Le Roy and Ferdinand Berthoud studied the conditions of isochronism pertaining to the spiral.

AN INVENTION THAT CREATED MUCH ENTHUSIASM.

However that may be, this magnificent invention, like the adaptation of the pendulum, was welcomed with general enthusiasm throughout the scientific world: without spiral and without pendulum, no other escapement but the recoil escapement was possible; a new highway was thus opened to the searchers. The water clocks (clepsy-

191

dræ) and the hour glasses disappeared completely, and the timepieces which had till then only marked the hours, having been perfected up to the point of keeping more exact time, were graced with the addition of another hand to tell off the minutes.

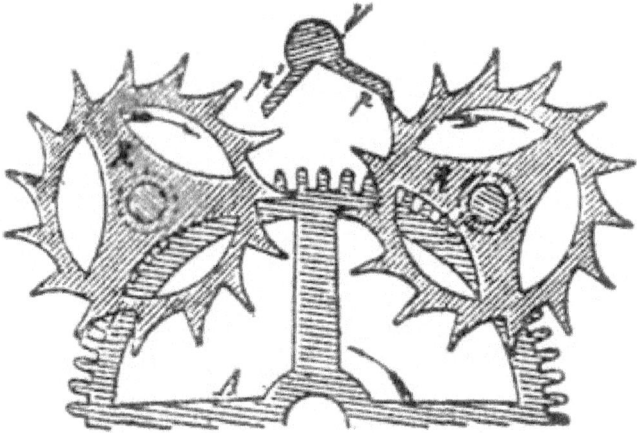

Fig. 160

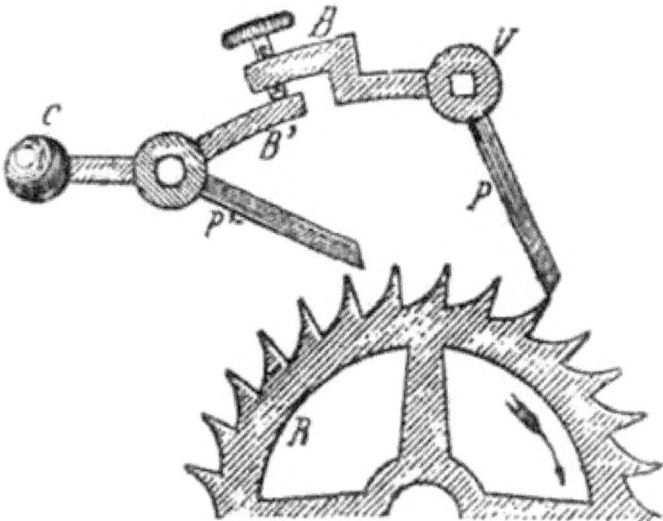

Fig. 161

It was not until 1695 that the first *dead-beat escapement* appeared upon the scene; during the interval of over twenty years all thought had been directed toward the one goal, viz.: the perfecting of the *verge escapement*; but practice demonstrated that no other arrangement of the parts was superior to the original idea. For the benefit of our readers we shall give a few of these attempts at betterment, and you may see for yourselves wherein the trials failed.

Fig. 157 represents a *verge escapement* with a ratchet wheel, the pallets *P P'* being carried upon separate axes. The two axes are rigidly connected, the one to the other, by means of the arms *o o'*. One of the axes carries besides the fork *F*, which transmits the impulse to the pendulum *B*. In the front view, at the right of the plate, for the sake of clearness the fork and the pendulum are not shown, but one may easily see the jointure of the arms *o o'* and their mode of operation.

Fig. 162

Another very peculiar arrangement of the *verge escapement* we show at Fig. 158. In this there are two wheels, one, *R'*, a small one in the form of a ratchet; the other, *R*, somewhat larger, called the bal-

193

ance wheel, but being supplied with straight and slender teeth. The verge *V* carrying the two pallets is pivoted in the vertical diameter of the larger wheel. The front view shows the *modus operandi* of this combination, which is practically the same as the others. The tooth *a* of the large wheel exerts its force upon the pallet *P*, and the tooth *b* of the ratchet will encounter the pallet *P'*. This pallet, after suffering its recoil, will receive the impulse communicated by the tooth *b*. This escapement surely could not have given much satisfaction, for it offers no advantage over the others, besides it is of very difficult construction.

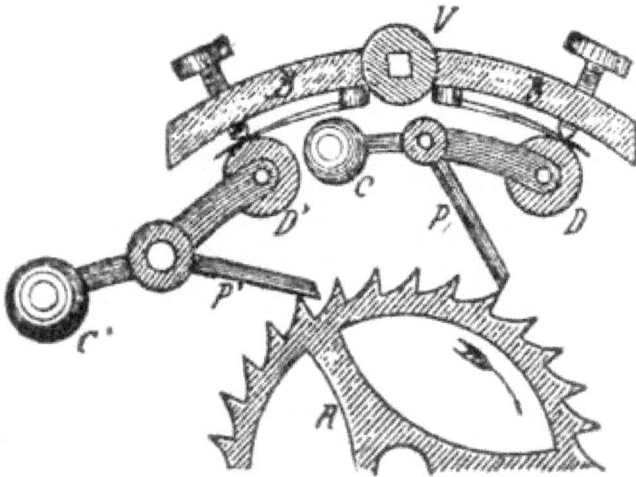

Fig. 163

INGENIOUS ATTEMPTS AT SOLUTION OF A DIFFICULT PROBLEM.

Much ingenuity to a worthy end, but of little practical value, is displayed in these various attempts at the solution of a very difficult problem. In Fig. 159 we have a mechanism combining two escape wheels engaging each other in gear; of the two wheels, *R R'*, one alone is driven directly by the train, the other being turned in the opposite direction by its comrade. Both are furnished with pins *c c'*, which act alternately upon the pallets *P P'* disposed in the same plane upon the verge *V* and pivoted between the wheels. Our drawing represents the escapement at the moment when the pin *C'* delivers its impulse, and this having been accomplished, the locking takes place upon the pin *C* of the other wheel upon the pallet *P'*. Another system of two escape wheels is shown in Fig. 160, but in this case the two wheels *R R* are driven in a like direction by the last wheel *A* of the train. The operation of the escapement is the same as in Fig. 159.

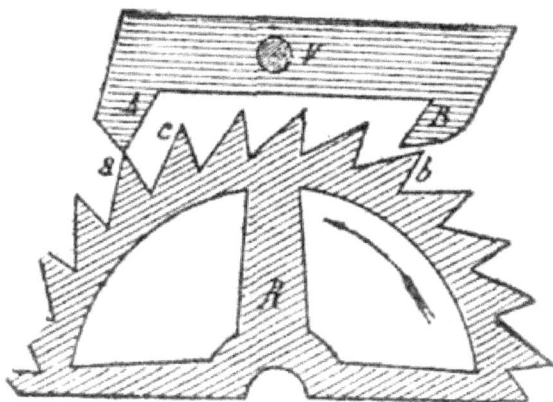

Fig. 164

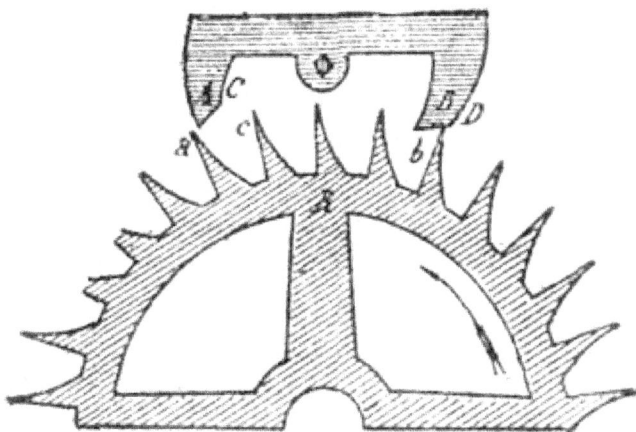

Fig. 165

195

In Fig. 161 we have a departure from the road ordinarily pursued. Here we see an escapement combining two levers, invented by the Chevalier de Béthune and applied by M. Thiout, master-horologist, at Paris in 1727. *P P'* are the two levers or pallets separately pivoted. Upon the axis *V*, of the lever *P*, is fixed a fork which communicates the motion to the pendulum. The two levers are intimately connected by the two arms *B B'*, of which the former carries an adjusting screw, a well-conceived addition for regulating the opening between the pallets. The counter-weight *C* compels constant contact between the arms *B B'*. The function is always the same, the recoil and the impulsion operate upon the two pallets simultaneously. This escapement enjoyed a certain degree of success, having been employed by a number of horologists who modified it in various ways.

VARIOUS MODIFICATIONS

Some of these modifications we shall show. For the first example, then, let Fig. 162 illustrate. In this arrangement the fork is carried upon the axis of the pallet *P'*, which effectually does away with the counter-weight *C*, as shown. Somewhat more complicated, but of the same intrinsic nature, is the arrangement displayed in Fig. 163. We should not imagine that it enjoyed a very extensive application. Here the two levers are completely independent of each other; they act upon the piece *B B* upon the axis *V* of the fork. The counter-weights *C C'* maintain the arms carrying the rollers *D D'* in contact with the piece *B B'* which thus receives the impulse from the wheel *R*. Two adjusting screws serve to place the escapement upon the center. By degrees these fantastic constructions were abandoned to make way for the anchor recoil escapement, which was invented, as we have said, in 1675, by G. Clement, a horologist, of London. In Fig. 164 we have the disposition of the parts as first arranged by this artist. Here the pallets are replaced by the inclines *A* and *B* of the anchor, which is pivoted at *V* upon an axis to which is fixed also the fork. The tooth *a* escapes from the incline or lever *A*, and the tooth *b* immediately rests upon the lever *B*; by the action of the pendulum the escape wheel suffers a recoil as in the pallet escapement, and on the return of the pendulum the tooth *c* gives out its impulse in the contrary direction. With this new system it became possible to increase the weight of the bob and at the same time lessen the effective motor power. The travel of the pendulum, or arc of oscillation, being reduced in a marked degree, an accuracy of rate was obtained far superior to that of the crown-wheel escapement. However, this new application of the recoil escapement was not adopted in France until 1695.

Fig. 166

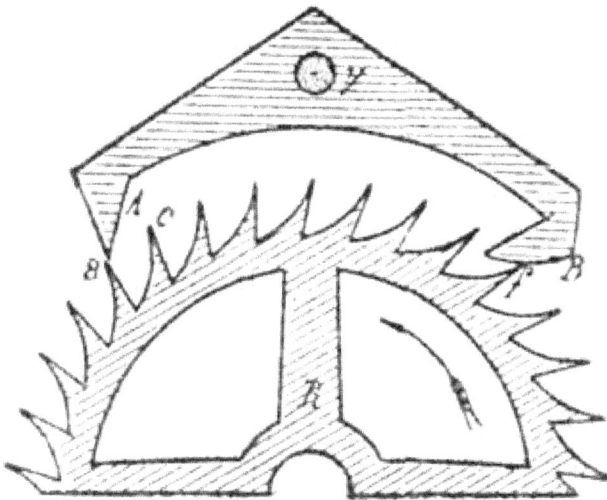

Fig. 167

The travel of the pendulum, though greatly reduced, still surpassed in breadth the arc in which it is isochronous, and repeated efforts were made to give such shape to the levers as would compel its oscillation within the arc of equal time; a motion which is, as was recognized even at that epoch, the prime requisite to a precise rating. Thus, in 1720, Julien Leroy occupied himself working out the proper shapes for the inclines to produce this desired isochronism. Searching along the same path, Ferd. Berthoud constructed an escapement represented by the Fig. 165. In it we see the same inclines *A B* of the former construction, but the locking is effected against the slides *C* and *D*, the curved faces of which produce isochronous oscillations of the pendulum. The tooth *b* imparts its lift and the tooth *c* will lock against the face *C*; after having passed through its recoil motion this tooth *c* will butt against the incline *A* and work out its lift or impulse upon it.

THE GABLE ESCAPEMENT.

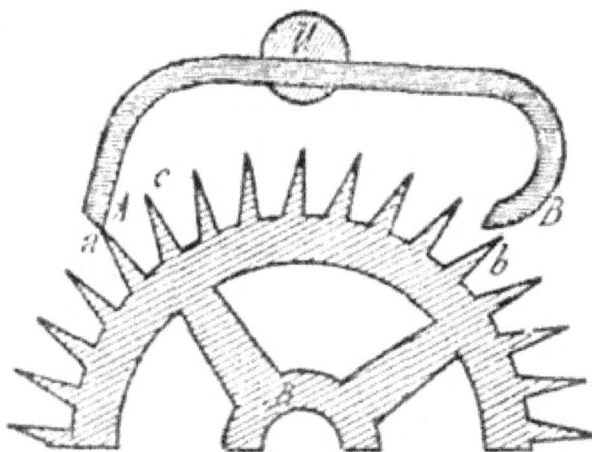

Fig. 168

198

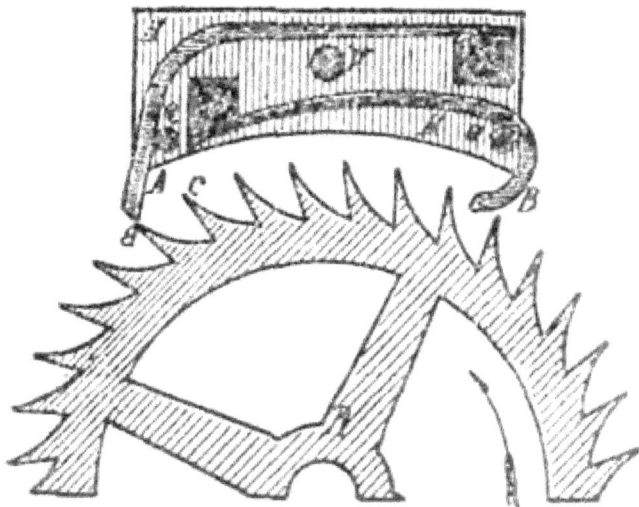

Fig. 169

The *gable escapement,* shown in Fig. 166, allows the use of a heavier pendulum, at the same time the anchor embraces within its jaws a greater number of the escape-wheel teeth; an arrangement after this manner leads to the conclusion that with these long levers of the anchor the friction will be considerably increased and the recoil faces will, as a consequence, be quickly worn away. Without doubt, this was invented to permit of opening and closing the contact points of the anchor more easily. Under the name of the *English recoil anchor* there came into use an escapement with a *reduced gable,* which embraced fewer teeth between the pallets or inclines; we give a representation of this in Fig. 167. This system seems to have been moderately successful. The anchor recoil escapement in use in Germany to-day is demonstrated in Fig. 168; this arrangement is also found in the American clocks. As we see, the anchor is composed of a single piece of curved steel bent to the desired curves. Clocks provided with this escapement keep reasonably good time; the resistance of the recoils compensate in a measure for the want of isochronism in the oscillations of the pendulum. Ordinary clocks require considerably more power to drive them than finer clocks and, as a consequence, their ticking is very noisy. Several means have been employed to dampen this noise, one of which we show in Fig. 169.

Fig. 170

Here the anchor is composed of two pieces, *A B*, screwed upon a plate *H* pivoting at *V*. In their arrangement the two pieces represent, as to distance and curvature, the counterpart of Fig. 168. At the moment of impact their extreme ends recoil or spring back from the shock of the escape teeth, but the resiliency of the metal is calculated to be strong enough to return them immediately to the contact studs *e e*.

As a termination to this chapter, we shall mention the use made at the present day of the recoil lever escapement in repeating watches. We give a diagram of this construction in Fig. 170. The lever here is intended to restrain and regulate the motion of the small striking work. It is pivoted at *V* and is capable of a very rapid oscillatory motion, the arc of which may, however, be fixed by the stud or stop *D*, which limits the swing of the fly *C*. This fly is of one piece with the lever and, together with the stud *D*, determines the angular motion of the lever. If the angle be large that means the path of the fly be long, then the striking train will move slowly; but if the teeth of the escape wheel *R* can just pass by without causing the lever to describe a supplementary or extended arc, the striking work will run off rapidly.

V. PUTTING IN A NEW CYLINDER

Putting in a new cylinder is something most watchmakers fancy they can do, and do well; but still it is a job very few workmen can do and fulfill all the requirements a job of this kind demands under the ever-varying conditions and circumstances presented in repairs of this kind. It is well to explain somewhat at this point: Suppose we have five watches taken in with broken cylinders. Out of this number probably two could be pivoted to advantage and make the watches as good as ever. As to the pivoting of a cylinder, we will deal with this later on. The first thing to do is to make an examination of the cylinder, not only to see if it is broken, but also to determine if pivoting is going to bring it out all right. Let us imagine that some workman has, at some previous time, put in a new cylinder, and instead of putting in one of the proper size he has put one in too large or too small. Now, in either case he would have to remove a portion of the escape-wheel tooth, that is, shorten the tooth: because, if the cylinder was too large it would not go in between the teeth, and consequently the teeth would have to be cut or stoned away. If the cylinder was too small, again the teeth would have to be cut away to allow them to enter the cylinder. All workmen have traditions, rules some call them, that they go by in relation to the right way to dress a cylinder tooth; some insisting that the toe or point of the tooth is the only place which should be tampered with. Other workmen insist that the heel of the tooth is the proper place. Now, with all due consideration, we would say that in ninety-nine cases out of a hundred the proper thing to do is to let the escape-wheel teeth entirely alone. As we can understand, after a moment's thought, that it is impossible to have the teeth of the escape wheel too long and have the watch run at all; hence, the idea of stoning a cylinder escape-wheel tooth should not be tolerated.

ESCAPE-WHEEL TEETH *vs.* CYLINDER.

It will not do, however, to accept, and take it for granted that the escape-wheel teeth are all right, because in many instances they have been stoned away and made too short; but if we accept this condition as being the case, that is, that the escape-wheel teeth are too short, what is the workman going to do about it? The owner of the watch will not pay for a new escape wheel as well as a new cylinder. The situation can be summed up about in this way, that we will have to make the best we can out of a bad job, and pick out and fit a cylinder on a compromise idea.

In regard to picking out a new cylinder, it may not do to select one of the same size as the old one, from the fact that the old one may not have been of the proper size for the escape wheel, because, even in new, cheap watches, the workmen who "run in" the escapement knew very well the cylinder and escape wheel were not adapted for each other, but they were the best he had. Chapter II, on the cylinder escapement, will enable our readers to master the subject and hence be better able to judge of allowances to be made in order to permit imperfect material to be used.

In illustration, let us imagine that we have to put in a new cylinder, and we have none of precisely the proper size, but we have them both a mere trifle too large and too small, and the question is which to use. Our advice is to use the smaller one if it does not require the escape-wheel teeth to be "dressed," that is, made smaller. Why we make this choice is based on the fact that the smaller cylinder shell gives less friction, and the loss from "drop"—that is, side play between the escape-wheel teeth and the cylinder—will be the same in both instances except to change the lost motion from inside to outside drop.

In devising a system to be applied to selecting a new cylinder, we meet the same troubles encountered throughout all watchmakers' repair work, and chief among these are good and convenient measuring tools. But even with perfect measuring tools we would have to exercise good judgment, as just explained. In Chapter II we gave a rule for determining the outside diameter of a cylinder from the diameter of the escape wheel; but such rules and tables will, in nine instances out of ten, have to be modified by attendant circumstances—as, for instance, the thickness of the shell of the cylinder, which should be one-tenth of the outer diameter of the shell, but the shell is usually thicker. A tolerably safe practical rule and one also depending very much on the workman's good judgment is, when the escape-wheel teeth have been shortened, to select a cylinder giving ample clearance inside the shell to the tooth, but by no means large enough to fill the space between the teeth. After studying carefully the instructions just given we think the workman will have no difficulty in selecting a cylinder of the right diameter.

MEASURING THE HEIGHTS.

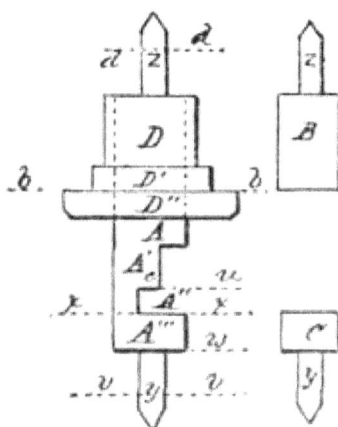

Fig. 171

The next thing is to get the proper heights. This is much more easily arrived at: the main measurement being to have the teeth of the escape wheel clear the upper face of the lower plug. In order to talk intelligently we will make a drawing of a cylinder and agree on the proper names for the several parts to be used in this chapter. Such drawing is shown at Fig. 171. The names are: The hollow cylinder, made up of the parts A A' A'' A''', called the shell—A is the great shell, A' the half shell, A'' the banking slot, and A''' the small shell. The brass part D is called the collet and consists of three parts—the hairspring seat D, the balance seat D' and the shoulder D'', against which the balance is riveted.

The first measurement for fitting a new cylinder is to determine the height of the lower plug face, which corresponds to the line x x, Fig. 171. The height of this face is such as to permit the escape wheel to pass freely over it. In selecting a new cylinder it is well to choose one which is as wide at the banking slot A'' as is consistent with safety. The width of the banking slot is represented by the dotted lines x u. The dotted line v represents the length to which the lower pivot y is to be cut.

Fig. 172

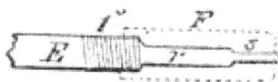

Fig. 173

There are several little tools on the market used for making the necessary measurements, but we will describe a very simple one which can readily be made. To do so, take about a No. 5 sewing needle and, after annealing, cut a screw thread on it, as shown at Fig, 172, where E represents the needle and t t the screw cut upon it. After the screw is cut, the needle is again hardened and tempered to a spring temper and a long, thin pivot turned upon it. The needle is now shaped as shown at Fig. 173. The pivot at s should be small enough to go easily through the smallest hole jewel to be found in cylinder watches, and should be about 1/16" long. The part at r should be about 3/16" long and only reduced in size enough to fully remove the screw threads shown at t.

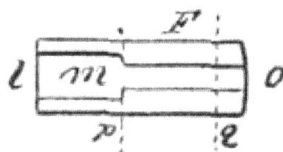

Fig. 174

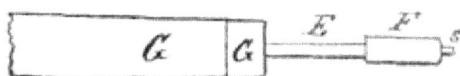

Fig. 175

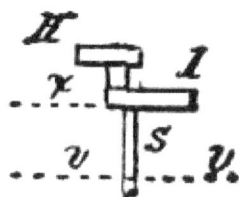

Fig. 176

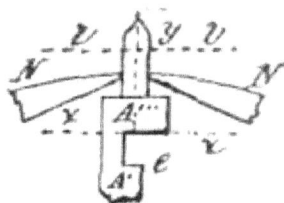

Fig. 177

We next provide a sleeve or guard for our gage. To do this we take a piece of hard brass bushing wire about 1/2" long and, placing it in a wire chuck, center and drill it nearly the entire length, leaving, say, 1/10" at one end to be carried through with a small drill. We show at F, Fig. 174, a magnified longitudinal section of such a sleeve. The piece F is drilled from the end l up to the line q with a drill of such a size that a female screw can be cut in it to fit the screw on the needle, and F is tapped out to fit such a screw from l up to the dotted line p. The sleeve F is run on the screw t and now appears as shown at Fig. 175, with the addition of a handle shown at G G'. It is evident that we can allow the pivot s to protrude from the sleeve F any portion of its length, and regulate such protrusion by the screw t. To employ this tool for getting the proper length to which to cut the pivot y, Fig. 171, we remove the lower cap jewel to the cylinder pivot and, holding, the movement in the left hand, pass the pivot s, Fig. 175, up through the hole jewel, regulate the length by turning the sleeve F until the arm of the escape wheel I, Fig. 176, will just turn free over it. Now the length of the pivot s, which protrudes beyond the sleeve F, coincides with the length to which we must cut the pivot y, Fig. 171. To hold a cylinder for reducing the length of the pivot y, we hold said pivot in a pair of thin-edged cutting pliers, as shown at Fig. 177, where $N N'$ represent the jaws of a pair of cutting pliers and y the pivot to be cut. The measurement is made by putting the pivot s between the jaws $N N'$ as they hold the pivot. The cutting is done by simply filing back the pivot until of the right length.

TURNING THE PIVOTS.

We have now the pivot y of the proper length, and what remains to be done is to turn it to the right size. We do not think it advisable to try to use a split chuck, although we have seen workmen drive the shell $A\ A'''$ out of the collet D and then turn up the pivots $y\ z$ in said wire chuck. To our judgment there is but one chuck for turning pivots, and this is the cement chuck provided with all American lathes. Many workmen object to a cement chuck, but we think no man should lay claim to the name of watchmaker until he masters the mystery of the cement chuck. It is not such a very difficult matter, and the skill once acquired would not be parted with cheaply. One thing has served to put the wax or cement chuck into disfavor, and that is the abominable stuff sold by some material houses for lathe cement. The original cement, made and patented by James Bottum for his cement chuck, was made up of a rather complicated mixture; but all the substances really demanded in such cement are ultramarine blue and a good quality of shellac. These ingredients are compounded in the proportion of 8 parts of shellac and 1 part of ultramarine—all by weight.

HOW TO USE A CEMENT CHUCK.

The shellac is melted in an iron vessel, and the ultramarine added and stirred to incorporate the parts. Care should be observed not to burn the shellac. While warm, the melted mass is poured on to a cold slab of iron or stone, and while plastic made into sticks about 1/2" in diameter.

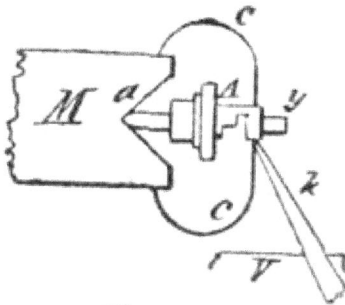

Fig. 178

Fig. 179

We show at Fig. 178 a side view of the outer end of a cement chuck with a cylinder in position. We commence to turn the lower pivot of a cylinder, allowing the pivot *z* to rest at the apex of the hollow cone *a*, as shown. There is something of a trick in turning such a hollow cone and leaving no "tit" or protuberance in the center, but it is important it should be done. A little practice will soon enable one to master the job. A graver for this purpose should be cut to rather an oblique point, as shown at *L*, Fig. 179. The slope of the sides to the recess *a*, Fig. 178, should be to about forty-five degrees, making the angle at *a* about ninety degrees. The only way to insure perfect accuracy of centering of a cylinder in a cement chuck is center by the shell, which is done by cutting a piece of pegwood to a wedge shape and letting it rest on the T-rest; then hold the edge of the pegwood to the cylinder as the lathe revolves and the cement soft and plastic. A cylinder so centered will be absolutely true. The outline curve at *c*, Fig. 178, represents the surface of the cement.

The next operation is turning the pivot to the proper size to fit the jewel. This is usually done by trial, that is, trying the pivot into the

207

hole in the jewel. A quicker way is to gage the hole jewel and then turn the pivot to the right size, as measured by micrometer calipers. In some cylinder watches the end stone stands at some distance from the outer surface of the hole jewel; consequently, if the measurement for the length of the pivot is taken by the tool shown at Fig. 175, the pivot will apparently be too short. When the lower end stone is removed we should take note if any allowance is to be made for such extra space. The trouble which would ensue from not providing for such extra end shake would be that the lower edge of the half shell, shown at e, Fig. 171, would strike the projection on which the "stalk" of the tooth is planted. After the lower pivot is turned to fit the jewel the cylinder is to be removed from the cement chuck and the upper part turned. The measurements to be looked to now are, first, the entire length of the cylinder, which is understood to be the entire distance between the inner faces of the two end stones, and corresponds to the distance between the lines v d, Fig. 171. This measurement can be got by removing both end stones and taking the distance with a Boley gage or a douzieme caliper.

A CONVENIENT TOOL FOR LENGTH MEASUREMENT.

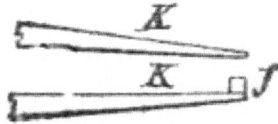

Fig. 180

A pair of common pinion calipers slightly modified makes as good a pair of calipers for length measurement as one can desire. This instrument is made by inserting a small screw in one of the blades— the head on the inner side, as shown at f, Fig. 180. The idea of the tool is, the screw head f rests in the sink of the cap jewel or end stone, while the other blade rests on the cock over the balance. After the adjusting screw to the caliper is set, the spring of the blades allows of their removal. The top pivot z of the cylinder is next cut to the proper length, as indicated by the space between the screwhead f and the other blade of the pinion caliper. The upper pinion z is held in the jaws of the cutting pliers, as shown in Fig. 177, the same as the lower one was

208

held, until the proper length between the lines *d v*, Fig. 171, is secured, after which the cylinder is put back into the cement chuck, as shown at Fig. 178, except this time the top portion of the cylinder is allowed to protrude so that we can turn the top pivot and the balance collet *D*, Fig. 171.

The sizes we have now to look to is to fit the pivot *z* to the top hole jewel in the cock, also the hairspring seat *D* and balance seat *D'*. These are turned to diameters, and are the most readily secured by the use of the micrometer calipers to be had of any large watchmakers' tool and supply house. In addition to the diameters named, we must get the proper height for the balance, which is represented by the dotted line *b*. The measurement for this can usually be obtained from the old cylinder by simply comparing it with the new one as it rests in the cement chuck. The true tool for such measurements is a height gage. We have made no mention of finishing and polishing the pivots, as these points are generally well understood by the trade.

REMOVING THE LATHE CEMENT.

One point perhaps we might well say a few words on, and this is in regard to removing the lathe cement. Such cement is usually removed by boiling in a copper dish with alcohol. But there are several objections to the practice. In the first place, it wastes a good deal of alcohol, and also leaves the work stained. We can accomplish this operation quicker, and save alcohol, by putting the cylinder with the wax on it in a very small homeopathic bottle and corking it tight. The bottle is then boiled in water, and in a few seconds the shellac is dissolved away. The balance to most cylinder watches is of red brass, and in some instances of low karat gold; in either case the balance should be repolished. To do this dip in a strong solution of cyanide of potassium dissolved in water; one-fourth ounce of cyanide in half pint of water is about the proper strength. Dip and rinse, then polish with a chamois buff and rouge.

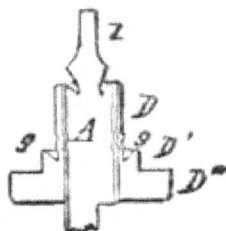

Fig. 181

In staking on the balance, care should be observed to set the banking pin in the rim so it will come right; this is usually secured by setting said pin so it stands opposite to the opening in the half shell. The seat of the balance on the collet D should be undercut so that there is only an edge to rivet down on the balance. This will be better understood by inspecting Fig. 181, where we show a vertical section of the collet D and cylinder A. At $g\ g$ is shown the undercut edge of the balance seat, which is folded over as the balance is rivetted fast.

About all that remains now to be done is to true up the balance and bring it to poise. The practice frequently adopted to poise a plain balance is to file it with a half-round file on the inside, in order not to show any detraction when looking at the outer edge of the rim. A better and quicker plan is to place the balance in a split chuck, and with a diamond or round-pointed tool scoop out a little piece of metal as the balance revolves. In doing this, the spindle of the lathe is turned by the hand grasping the pulley between the finger and thumb. The so-called diamond and round-pointed tools are shown at $o\ o'$, Fig. 182. The idea of this plan of reducing the weight of a balance is, one of the tools o is rested on the T-rest and pressed forward until a chip is started and allowed to enter until sufficient metal is engaged, then, by swinging down on the handle of the tool, the chip is taken out.

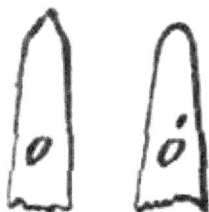

Fig. 182

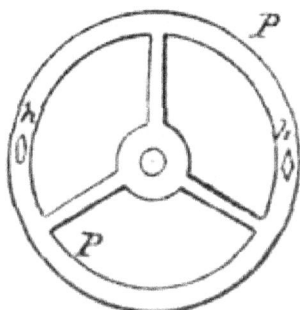

Fig. 183

In placing a balance in a step chuck, the banking pin is caused to enter one of the three slots in the chuck, so as not to be bent down on to the rim of the balance. It is seldom the depth between the cylinder and escape wheel will need be changed after putting in a new cylinder; if such is the case, however, move the chariot—we mean the cock attached to the lower plate. Do not attempt to change the depth by manipulating the balance cock. Fig. 183 shows, at *h h*, the form of chip taken out by the tool *o o'*, Fig. 182.

Also from Benediction Books ...

Wandering Between Two Worlds: Essays on Faith and Art
Anita Mathias
Benediction Books, 2007
152 pages
ISBN: 0955373700

Available from www.amazon.com, www.amazon.co.uk
www.wanderingbetweentwoworlds.com

In these wide-ranging lyrical essays, Anita Mathias writes, in lush, lovely prose, of her naughty Catholic childhood in Jamshedpur, India; her large, eccentric family in Mangalore, a sea-coast town converted by the Portuguese in the sixteenth century; her rebellion and atheism as a teenager in her Himalayan boarding school, run by German missionary nuns, St. Mary's Convent, Nainital; and her abrupt religious conversion after which she entered Mother Teresa's convent in Calcutta as a novice. Later rich, elegant essays explore the dualities of her life as a writer, mother, and Christian in the United States-- Domesticity and Art, Writing and Prayer, and the experience of being "an alien and stranger" as an immigrant in America, sensing the need for roots.

About the Author

Anita Mathias was born in India, has a B.A. and M.A. in English from Somerville College, Oxford University and an M.A. in Creative Writing from the Ohio State University. Her essays have been published in The Washington Post, The London Magazine, The Virginia Quarterly Review, Commonweal, Notre Dame Magazine, America, The Christian Century, Religion Online, The Southwest Review, Contemporary Literary Criticism, New Letters, The Journal, and two of HarperSanFrancisco's The Best Spiritual Writing anthologies. Her non-fiction has won fellowships from The National Endowment for the Arts; The Minnesota State Arts Board; The Jerome Foundation, The Vermont Studio Center; The Virginia Centre for the Creative Arts, and the First Prize for the Best General Interest Article from the Catholic Press Association of the United States and Canada. Anita has taught Creative Writing at the College of William and Mary, and now lives and writes in Oxford, England.

"Yesterday's Treasures for Today's Readers"

Titles by Benediction Classics available from Amazon.co.uk

Religio Medici, Hydriotaphia, Letter to a Friend, Thomas Browne

Pseudodoxia Epidemica: Or, Enquiries into Commonly Presumed Truths, Thomas Browne

The Maid's Tragedy, Beaumont and Fletcher

The Custom of the Country, Beaumont and Fletcher

Philaster Or Love Lies a Bleeding, Beaumont and Fletcher

A Treatise of Fishing with an Angle,   Dame Juliana Berners.

Pamphilia to Amphilanthus, Lady Mary Wroth

The Compleat Angler, Izaak Walton

The Magnetic Lady, Ben Jonson

Every Man Out of His Humour, Ben Jonson

The Masque of Blacknesse. The Masque of Beauty,. Ben Jonson

The Life of St. Thomas More, William Roper

Pendennis, William Makepeace Thackeray

Salmacis and Hermaphroditus attributed to Francis Beaumont

Friar Bacon and Friar Bungay Robert Greene

Holy Wisdom, Augustine Baker

The Jew of Malta and the Massacre at Paris, Christopher Marlowe

Tamburlaine the Great, Parts 1 & 2 AND Massacre at Paris, Christopher Marlowe

All Ovids Elegies, Lucans First Booke, Dido Queene of Carthage, Hero and Leander, Christopher Marlowe

The Titan, Theodore Dreiser

Scapegoats of the Empire: The true story of the Bushveldt Carbineers, George Witton

All Hallows' Eve, Charles Williams

The Place of The Lion, Charles Williams

The Greater Trumps, Charles Williams

My Apprenticeship: Volumes I and II, Beatrice Webb

Last and First Men / Star Maker, Olaf Stapledon

Last and First Men, Olaf Stapledon

Darkness and the Light, Olaf Stapledon

The Worst Journey in the World, Apsley Cherry-Garrard

The Schoole of Abuse, Containing a Pleasaunt Invective Against Poets, Pipers, Plaiers, Iesters and Such Like Catepillers of the Commonwelth, Stephen Gosson

Russia in the Shadows, H. G. Wells

Wild Swans at Coole, W. B. Yeats

A hundreth good pointes of husbandrie, Thomas Tusser

The Collected Works of Nathanael West: "The Day of the Locust", "The Dream Life of Balso Snell", "Miss Lonelyhearts", "A Cool Million", Nathanael West

Miss Lonelyhearts & The Day of the Locust, Nathaniel West

The Worst Journey in the World, Apsley Cherry-Garrard

Scott's Last Expedition, V1, R. F. Scott

The Dream of Gerontius, John Henry Newman

The Brother of Daphne, Dornford Yates

The Poetry of Architecture: Or the Architecture of the Nations of Europe Considered in Its Association with Natural Scenery and National Character, John Ruskin

The Downfall of Robert Earl of Huntington, Anthony Munday

Clayhanger, Arnold Bennett

South: The Story of Shackleton's Last Expedition 1914-1917, Sir Ernest Shackketon

Greene's Groatsworth of Wit: Bought With a Million of Repentance, Robert Greene

Beau Sabreur, Percival Christopher Wren

The Hekatompathia, or Passionate Centurie of Love, Thomas Watson

The Art of Rhetoric, Thomas Wilson

Stepping Heavenward, Elizabeth Prentiss

Barker's Delight, or The Art of Angling, Thomas Barker
The Napoleon of Notting Hill, G.K. Chesterton

The Douay-Rheims Bible (The Challoner Revision)

Endimion - The Man in the Moone, John Lyly

Gallathea and Midas, John Lyly,

Manners, Custom and Dress During the Middle Ages and During the Renaissance Period, Paul Lacroix

Obedience of a Christian Man, William Tyndale

St. Patrick for Ireland, James Shirley

The Wrongs of Woman; Or Maria/Memoirs of the Author of a Vindication of the Rights of Woman, Mary Wollstonecraft and William Godwin

De Adhaerendo Deo. Of Cleaving to God, Albertus Magnus

Obedience of a Christian Man, William Tyndale

A Trick to Catch the Old One, Thomas Middleton

A Yorkshire Tragedy, Thomas Middleton (attrib.)

The Princely Pleasures at Kenelworth Castle, George Gascoigne

The Fair Maid of the West. Part I and Part II. Thomas Heywood

Proserpina, Volume I and Volume II. Studies of Wayside Flowers, John Ruskin

The Endeavour Journal of Sir Joseph Banks. Sir Joseph Banks

Christ Legends: And Other Stories, Selma Lagerlof; (trans. Velma Swanston Howard)

Chamber Music, James Joyce

Blurt, Master Constable, Thomas Middleton, Thomas Dekker

Since Yesterday, Frederick Lewis Allen

The Scholemaster: Or, Plaine and Perfite Way of Teachyng Children the Latin Tong , Roger Ascham

The Wonderful Year, 1603, Thomas Dekker

Waverley, Sir Walter Scott

Guy Mannering, Sir Walter Scott

Old Mortality, Sir Walter Scott

The Knight of Malta, John Fletcher

Space Prison, Tom Godwin

The Home of the Blizzard Being the Story of the Australasian Antarctic Expedition, 1911-1914, Douglas Mawson

Wild-goose Chase , John Fletcher

If You Know Not Me, You Know Nobody. Part I and Part II, Thomas Heywood

The Ragged Trousered Philanthropists, Robert Tressell

The Island of Sheep, John Buchan

Eyes of the Woods, Joseph Altsheler

The Club of Queer Trades, G. K. Chesterton

The Financier, Theodore Dreiser

Something of Myself, Rudyard Kipling

Law of Freedom in a Platform, or True Magistracy Restored, Gerrard Winstanley

Damon and Pithias, Richard Edwards

Dido Queen of Carthage: And, The Massacre at Paris, Christopher Marlowe

Cocoa and Chocolate: Their History from Plantation to Consumer, Arthur Knapp

Lady of Pleasure, James Shirley

The South Pole: An account of the Norwegian Antarctic expedition in the "Fram," 1910-12. Volume 1 and Volume 2, Roald Amundsen

A Yorkshire Tragedy, Thomas Middleton (attrib.)

The Tragedy of Soliman and Perseda, Thomas Kyd

The Rape of Lucrece. Thomas Heywood

Myths and Legends of Ancient Greece and Rome, E. M. Berens

and many others…

Tell us what you would love to see in print again, at affordable prices!
Email: **benedictionbooks@btinternet.com**

9 781849 020343